Topics in Current Chemistry C

Aims and Scope

The series Topics in Current Chemistry Collections presents critical reviews from the journal Topics in Current Chemistry organized in topical volumes. The scope of coverage is all areas of chemical science including the interfaces with related disciplines such as biology, medicine and materials science.

The goal of each thematic volume is to give the non-specialist reader, whether in academia or industry, a comprehensive insight into an area where new research is emerging which is of interest to a larger scientific audience.

Each review within the volume critically surveys one aspect of that topic and places it within the context of the volume as a whole. The most significant developments of the last 5 to 10 years are presented using selected examples to illustrate the principles discussed. The coverage is not intended to be an exhaustive summary of the field or include large quantities of data, but should rather be conceptual, concentrating on the methodological thinking that will allow the non-specialist reader to understand the information presented.

Contributions also offer an outlook on potential future developments in the field.

More information about this series at http://www.springer.com/series/14181

Milan Vrabel • Thomas Carell
Editors

Cycloadditions in Bioorthogonal Chemistry

With contributions from

T. Carell • P. R. Chen • F. L. van Delft • N. K. Devaraj
J. Dommerholt • A. Herner • S. Kath-Schorr • Q. Lin
F. P. J. T. Rutjes • M. Vrabel • H. Wu • M. Yang • Y. Yang

Springer

Editors
Milan Vrabel
Department of Bioorganic
and Medicinal Chemistry
Institute of Organic Chemistry
and Biochemistry
Prague, Czech Republic

Thomas Carell
Department of Chemistry
and Pharmacy
Ludwig-Maximilians University Munich
Munich, Germany

Originally published in Top Curr Chem (Z) Volume 374 (2016),

ISSN 2367-4067 ISSN 2367-4075 (electronic)
Topics in Current Chemistry Collections
ISBN 978-3-319-80626-6 ISBN 978-3-319-29686-9 (eBook)
DOI 10.1007/978-3-319-29686-9

Printed on acid-free paper

This Springer imprint is published by Springer Nature
The registered company is Springer International Publishing AG Switzerland

Contents

Top Curr Chem (Z) (2016) 374:15
DOI 10.1007/s41061-016-0017-3

EDITORIAL

M. Vrabel and T. Carell for Cycloadditions in Bioorthogonal Chemistry

T. Carell[1] · M. Vrabel[2] · M. Yang[1,2] · Y. Yang[1,2] · P. R. Chen[1,2] · J. Dommerholt[1,2] · F. P. J. T. Rutjes[1,2] · F. L. van Delft[1,2] · A. Herner[1,2] · Q. Lin[1,2] · H. Wu[1,2] · N. K. Devaraj[1,2] · S. Kath-Schorr[1,2]

Received: 21 January 2016 / Accepted: 19 February 2016 / Published online: 16 March 2016

1 Aims and Scope

The series Topics in Current Chemistry presents critical reviews of the present and future trends in modern chemical research. The scope of coverage is all areas of chemical science, including the interfaces with related disciplines, such as biology, medicine and materials science.

The goal of each thematic volume is to give the non-specialist reader, whether in academia or industry, a comprehensive insight into an area where new research is emerging that is of interest to a larger scientific audience.

Each review within the volume critically surveys one aspect of that topic and places it within the context of the volume as a whole. The most significant developments of the last 5–10 years are presented using selected examples to illustrate the principles discussed. The coverage is not intended to be an exhaustive summary of the field or to include large quantities of data, but should rather be conceptual, concentrating on the methodological thinking that will allow the non-specialist reader to understand the information presented. Contributions also offer an outlook on potential future developments in the field.

This article is part of the Topical Collection "Cycloadditions in Bioorthogonal Chemistry"; edited by Milan Vrabel and Thomas Carell.

✉ T. Carell
Thomas.carell@cup.uni-muenchen.de

✉ M. Vrabel
vrabel@uochb.cas.cz

1 Ludwig-Maximilians University, Butenandtstrasse 5-13, Building F, 81377 Munich, Germany

2 Institute of Organic Chemistry and Biochemistry, AS CR, v.v.i., Flemingovo nam. 2, Building B, 16610 Prague, Czech Republic

Review articles for the individual volumes are invited by the volume editors.

2 Preface

Due to fantastic progress in the art of chemical synthesis, organic chemists today have access to a great arsenal of powerful chemical transformations that enable the synthesis of even the most complex structures and molecules. While some of the most reliable transformation have been known for decades, the repertoire of new, broadly applicable reactions is still expanding, so that even chemical transformations directly on biomolecules have now become routinely possible. Today, high yielding and efficient chemical reactions on biomolecules such as nucleic acids, proteins or oligosaccharides are often summarized under the term bioorthogonal click reactions. In this field, chemists are becoming part of interdisciplinary research groups in which they work hand in hand with biochemists, enzymologists and researchers in the discipline of biophysics.

Bioorthogonal chemistry uses chemical transformations that can be performed in the presence of all types of biomolecules and even in living cells, and which occur between two reactive groups that do not react with all the functional groups present on the biomolecules. Given the complexity of living matter, this is a formidable achievement. Once one of the reactive groups is incorporated into the biomolecule of interest, the biomolecule can be selective modified at the given position, even inside a living cell. This would have been impossible 15 years ago.

The development of bioorthogonal chemical transformations goes hand in hand with the invention of new strategies that allow the selective incorporation of bioorthogonal reactive groups into biomolecules. For nucleic acids, this involves the development of phosphoramidite building blocks and special triphosphates. The oligonucleotides are then constructed by solid phase chemistry or enzymatically using the PCR reaction. For proteins, new tools based on the suppression of stop codons are now available, allowing the incorporation of unnatural amino acids into proteins. These amino acids contain bioorthogonal groups at their side chains. For adding bioorthogonal groups into oligosaccharides, one typically uses feeding experiments with modified sugars, which are then inserted into the oligosaccharide using biosynthetic pathways.

This book, in its content, is a unique collection of contributions, compiling the huge progress that has been made in the use of cycloaddition reactions in bioorthogonal chemistry. Among other chemical transformations that are currently routinely used for bioconjugations, the cycloaddition reactions stand out as the most versatile and powerful methods. Based on the pioneering work of chemists such as Rolf Huisgen, Kurt Alder and Otto Diels, who worked out the mechanisms of these transformations decades ago, a new generation of chemists has started to adapt these reactions for manipulation and tagging of biomolecules.

In this setting, this book is therefore also a tribute to these scientific heroes, who until today dominate large parts of our chemistry textbooks and who we all know so well from our studies.

Individual chapters provide overviews on cycloaddition reactions used for biomolecule labeling—the first part of the book deals with dipolar cycloadditions, while the last two chapters are devoted to Diels–Alder-type reactions. The different aspects of the chemistry are outlined and the benefits of each methodology are demonstrated by discussing specific applications and possible future directions in the field. We are deeply indebted to all the authors for their willingness to be a part of this interesting project and for sharing their knowledge and experience in this rapidly growing field. With this book, we hope to invoke interest among students and colleagues in bioorthogonal chemistry, and to encourage and enthuse a new generation of scientists in this exciting scientific area.

Milan Vrabel

Thomas Carell

Top Curr Chem (Z) (2016) 374:9
DOI 10.1007/s41061-016-0010-x

REVIEW

Bioorthogonal Chemistry—Introduction and Overview

Thomas Carell[1] · **Milan Vrabel**[2]

Received: 26 October 2015 / Accepted: 15 January 2016 / Published online: 1 February 2016

Abstract Bioorthogonal chemistry has emerged as a new powerful tool that facilitates the study of structure and function of biomolecules in their native environment. A wide variety of bioorthogonal reactions that can proceed selectively and efficiently under physiologically relevant conditions are now available. The common features of these chemical reactions include: fast kinetics, tolerance to aqueous environment, high selectivity and compatibility with naturally occurring functional groups. The design and development of new chemical transformations in this direction is an important step to meet the growing demands of chemical biology. This chapter aims to introduce the reader to the field by providing an overview on general principles and strategies used in bioorthogonal chemistry. Special emphasis is given to cycloaddition reactions, namely to 1,3-dipolar cycloadditions and Diels–Alder reactions, as chemical transformations that play a predominant role in modern bioconjugation chemistry. The recent advances have established these reactions as an invaluable tool in modern bioorthogonal chemistry. The key aspects of the methodology as well as future outlooks in the field are discussed.

Keywords Bioorthogonal reactions · Click chemistry · Biomolecule labeling · 1,3-dipolar cycloaddition · Diels–Alder reaction

✉ Milan Vrabel
vrabel@uochb.cas.cz

Thomas Carell
thomas.carell@cup.uni-muenchen.de

1 Department of Chemistry and Pharmacy, Ludwig-Maximilians University Munich, Butenandtstrasse 5–13, 81377 Munich, Germany

2 Department of Bioorganic and Medicinal Chemistry, Institute of Organic Chemistry and Biochemistry, Academy of Sciences of the Czech Republic, Flemingovo nam. 2, 16610 Prague 6, Czech Republic

1 Introduction

Over the centuries, chemists have developed a plethora of chemical transformations that enable the synthesis of complex molecules in a straightforward and efficient manner. Typical chemical reactions in organic chemistry are performed in organic solvents over a wide range of temperatures, pressures, and pH. Often, the use of protecting groups must also be employed to prevent side-reactions and the formation of undesired by-products. In addition, the inherent instability of many reagents often necessitates performing the reaction under strictly anhydrous conditions and under an inert atmosphere devoid of oxygen. In stark contrast, in nature, biochemical reactions proceed in an aqueous environment under physiological conditions that for most organisms are at near-neutral pH, moderate temperature, atmospheric pressure, and in the presence of oxygen. Nature has developed sophisticated ways to carry out vital biosyntheses under these conditions. Enzymes, the catalysts of living organisms, are fantastic instruments of evolution able to precisely control chemical reactions with imposing selectivity and efficiency. While this works extraordinarily well in nature, the development of new artificial enzymes that would enable the covalent labeling of desired biomolecules is impractical for scientists. The ever-growing desire to understand the numerous biological processes taking place in living organisms fueled the development of a new strategy based on pure chemical reactions that can proceed under physiological conditions without perturbing the structure and activity of the delicate biomolecules. Although this might look trivial, in fact, this represents a considerable challenge for organic chemists since such chemistry must fulfill several criteria. The reaction must work in aqueous media and cannot interfere with the myriad of functionalities found in vivo. Due to the reducing environment in the cytosol, the reaction cannot be sensitive to redox chemistry. In addition, the reaction must proceed with reasonable kinetics, because biomolecules in cells are present only in low concentrations. The introduction of the click chemistry concept by Sharpless [1] inspired many organic chemists throughout the world to start work intensively on developing chemical reactions that fulfill these criteria and are compatible with biological systems. There are a handful of bioorthogonal reactions currently available that significantly extend our ability to label, study, and manipulate biomolecules.

In this book, we provide a general overview on bioorthogonal chemical reactions with a focus on cycloaddition reactions. The 1,3-dipolar cycloadditions and Diels–Alder reactions are extremely versatile chemical reactions and have been known in organic chemistry for decades. However, it was only recently that chemical biologists realized the potential of these transformations and started to employ them for bioconjugation strategies. A significant amount of work has been done to boost the kinetics, to optimize the reagents and to improve the biocompatibility of these reactions. Recent advances have enabled the development of this chemistry into a robust tool that is routinely utilized by scientists around the world for applications ranging from synthetic chemistry and biology to polymer sciences and drug discovery. The individual chapters of this book aim to introduce 1,3-dipolar cycloadditions and Diels–Alder reactions in connection with their use as bioorthogonal ligation methods. Emphasis is concentrated on the progress made in recent years and also on future aspects of the methodologies in modern bioorthogonal chemistry.

2 Principles of Bioorthogonal Reactions

Our understanding of biological processes directly depends on our ability to visualize and manipulate the key biomolecules involved in them. The development of genetically encoded fluorescent proteins (e.g. green fluorescent protein, GFP) has revolutionized the tracking and detection of native proteins in living systems [2, 3]. The huge impact of this methodology was recognized with the 2008 Nobel Prize in Chemistry [4]. However, other biomolecules, such as lipids, glycans and nucleic acids, are not amenable to these genetically encoded reporters. A more general platform is therefore needed to tag these biomolecules. One such strategy relies on the installation of a special functional group of a non-natural origin into the target biomolecule. This functionality equips the intended biomolecule with a unique chemical handle that is distinct from all the other functionalities found in nature and that can react, ideally, only with a second appropriate orthogonal reaction partner (Fig. 1).

The term bioorthogonal chemical reaction was originally coined by Bertozzi [5]. It refers to a chemical reaction that neither interacts nor interferes with biological systems. Historically, the concept of bioorthogonal reactions builds on much older methodologies that utilize unique reactivity of functional groups present on amino acid side chains (e.g. on lysine and cysteine). Although this type of chemistry still belongs to most popular bioconjugation techniques, the fact that these functionalities occur in nature in multiple copies limits their use for in vitro applications. In order to tag biomolecules in their truly native environment, e.g. in living cells or whole organism (in vivo), it is necessary to find functionalities that are not so prevalent among (or ideally are absent from) biomolecules. Already in 1990, the Rideout group described the successful assembly of hydrazone-containing toxins from aldehyde and acylhydrazine precursors within live cells [6]. Later on, in 1998,

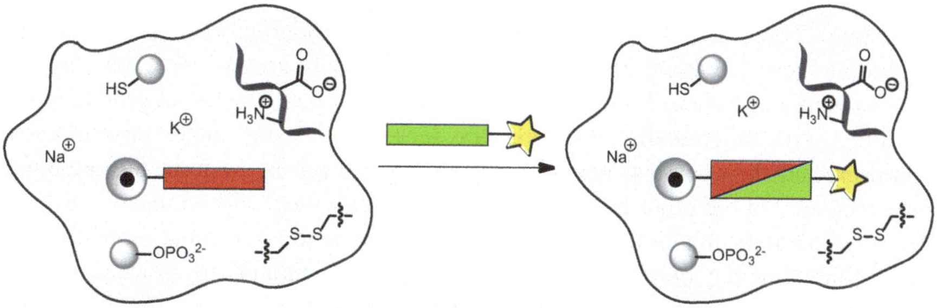

Fig. 1 A general bioorthogonal chemical reaction scheme. Two bioorthogonal functional groups (shown as *rectangles*) react selectively with each other in the presence of other naturally occurring functionalities

Tsien and coworkers showed that tetra-cysteine motif embedded into protein sequence can be used to tag proteins inside live cells with fluorogenic bisarsenical dyes [7]. These pioneering studies inspired organic chemists throughout the world, who started to explore and re-design various chemical reactions to proceed selectively under physiological conditions and even in living organisms [8].

The whole process of designing and developing new bioorthogonal reactions usually starts with intensive work in an organic chemistry laboratory where the initial experiments are done in a flask. After demonstrated success under chemical conditions, more physiological-like conditions are then employed in order to verify that the chemistry is compatible with biology. Although this first stage is primarily done by synthetic organic chemists, the second part of the process, when the respective functional group must be installed on biomolecules, often requires assistance of someone with a more biological background. This is also true for the next phase, when the chemistry is used to visualize the studied biomolecule or in order to answer a specific biological problem. As it is also necessary to perform careful analysis of the data and individual steps throughout the experiment, it soon becomes clear that this is a multidisciplinary field where close collaboration between scientists with different expertise is crucial for the final outcome. This research area is an elegant example showcasing joint effort among scientists to provide new powerful tools that exceed the boundaries of individual scientific disciplines.

2.1 Design of Bioorthogonal Functional Groups

Recent developments in chemical biology have led to the discovery of a number of reactive functional groups having features suitable for bioconjugation. It is worthy of mention that the bulk of modern bioconjugation reactions were not truly newly discovered. Instead, well-known and established chemical reactions were rediscovered and optimized in order to proceed selectively and efficiently in a biological environment. This is not meant to underestimate the amount of work that has been done in the area, but rather to acknowledge the great job of previous chemists who discovered the chemistry, often decades ago. There is no doubt that the success of modern bioconjugation reactions would not be possible without significant recent contributions of researchers active in the field.

The design of chemical functional groups suitable for bioconjugation is not a trivial task. There are many factors that must be considered when designing and developing new chemical ligation methods [9]. Such groups must be inert to aqueous media and should not react or interfere with other functional groups found in Nature. The functionality must also be reasonably stable under physiological conditions and the chemical reagents utilized should not be toxic to the organism being studied. On the other hand, the functional group must still be reactive toward the complementary functional group that participates in the bioconjugation reaction. Ideally, the kinetics of the reaction should be fast enough to proceed at low concentrations and in a reasonable time frame as well. This can be particularly problematic, since high reactivity is often linked to reduced stability and a higher probability of side reactions occurring. Fast reaction rates also enable the use of nearly equimolar amounts of reagents, making the process economically favorable. The use of functional groups that can react with multiple reacting partners can also be beneficial in some cases. This allows for new possibilities in choosing the best reacting partner for the planned application. Correctly selecting the orthogonal functional groups also depends on the type of target biomolecule. The characteristics of a given functional groups found in each specific biomolecule usually directly determine its structure and function. This should be reflected in the type of the chemistry and thus the functional group to be inserted into the biomolecule. Moreover, it is important to consider the possible influence of the added unnatural functional group on the structure and function of the target biomolecule. The correct selection of a suitable site for the non-natural functional group within the biomolecule can be crucial in this regard. Small-sized functional groups are generally preferred over bulkier ones.

There are several bioorthogonal functional groups currently available. Each of these has its advantages, but also shortcomings. There is no perfect bioorthogonal reaction known to date, although different aspects of a particular chemical reaction make it more likely to be suitable for a certain application. The diversity in bioorthogonal functionality is reflected in the variety of applications for which they can be used. The enormous progress that has been made in developing new bioconjugation reactions within recent years offers us a whole new repertoire of chemistry that can be used for efficient and selective labeling of biomolecules. This provides us with great possibility and opportunity to study various biomolecules in their native environment.

2.2 Incorporation of Bioorthogonal Functional Groups into Biomolecules

The optimization of functional groups and the fine tuning of reaction conditions are an indisputable part of the process toward the development of a new bioconjugation reaction. Correct use of the developed methodology for tagging biomolecules with various probes requires that one of the reagents is somehow installed in the target biomolecule. There are several ways to do this, and we will now describe the most commonly employed strategies.

Obviously, the type of the biomolecule directly affects the way the functional group is introduced into it, because different biomolecules are made of different building blocks. One straightforward strategy commonly employed for introducing unnatural moieties to biomolecules is automated synthesis. Automated oligonucleotide and peptide synthesis enables the production of relatively long chains of oligomers containing both natural and unnatural building blocks. The sequence of the building blocks can be programmed, and thus the position of the unnatural functionality within the structure is not limited. The only limitations in the methodology are: the synthetic accessibility of the unnatural building blocks and the compatibility of the non-natural functional group with the chemistry used for the oligomer synthesis.

There are many known examples where modified phosphoramidites bearing various bioorthogonal functional groups have been used for the introduction of a ligation handle to nucleic acids [10]. Similarly, unnatural amino acid building blocks equipped with clickable functionality have been used for the synthesis of modified peptides [11]. The commercial availability of the building blocks makes this strategy a straightforward method for incorporating various useful functional moieties to biopolymers.

A different strategy is used for introducing these moieties to large biomolecules. These methods exploit the promiscuity of some natural enzymes. Several DNA polymerases are known to tolerate various structural analogs of their natural substrates. These enzymes are able to recognize unnatural nucleoside triphosphates as substrates and use them during the biosynthesis of a DNA strand. In this way, long DNA sequences containing diverse artificial functionalities can be prepared [12].

The most commonly employed strategy for producing modified proteins is based on the so-called amber suppression technology. Here, a naturally occurring amber stop codon (UAG) is reprogrammed to serve as a sense codon that is recognized by a specific tRNA molecule bearing an unnatural amino acid [13]. The unnatural amino acid is loaded onto the tRNA molecule by natural or evolved tRNA synthetase that is able to recognize the specific non-natural amino acid as substrate [14]. By genetic encoding of the UAG amber codon, it is possible to direct the desired modified amino acid to a particular position within the sequence of the target protein. This biosynthetic machinery can be incorporated into live cells (e.g. *E. coli*), which are then able to produce the site-specifically modified proteins containing the desired clickable functional group. This functionality can be then used for attaching various tags by the appropriate bioorthogonal reaction.

While these methodologies work well for nucleic acids and proteins, a different strategy is used to equip glycans and lipids with bioorthogonal chemical handles. This strategy is based on the metabolic incorporation of natural substrate analogs. These analogs are first made using common organic synthesis and are then simply added to the culture medium of growing cells. These metabolic precursors are taken up by the cells, where they are metabolized by the biosynthetic machinery, and, in a slightly modified form, are finally incorporated into the respective biomolecules. This strategy has been successfully applied for metabolic labeling of glycans [15] and phospholipids [16], but works for nucleic acids [17] and peptidoglycans [18] as well.

In the following sections, we introduce two types of chemical reactions that gained special attention in the field of bioorthogonal chemistry: the 1,3-dipolar cycloadditions and Diels–Alder reactions.

3 1,3-Dipolar Cycloadditions

The concept of (3 + 2)-dipolar cycloadditions, pioneered by Rolf Huisgen, dates back to the early 1960s [19]. Over the years, this reaction has found widespread utility in heterocyclic chemistry, natural product synthesis and medicinal chemistry [20–22]. These elegant chemical transformations facilitate the construction of products with high complexity where two new σ-bonds are formed in a single step. In a typical (3 + 2)-dipolar cycloaddition reaction, a 1,3-dipole reacts with a multiple bond system (usually an alkene or alkyne), termed the dipolarophile, giving rise to a new five membered ring. The dipole with the general formula *a-b-c* usually contains an atom *a* with an electron sextet, i.e. an incomplete valence shell combined with a positive formal charge, and atom *c*, the negatively charged center, which has an unshared electron pair. The 1,3-dipoles can be drawn using sextet or octet resonance structures (Fig. 2). Dipoles with a 'double' bond are typically referred to as propargylic species, while without the double bond, they are referred to as allenic species.

Studies on the mechanism of 1,3-dipolar cycloaddition reactions have shown that the reaction proceeds in a concerted fashion and is controlled not only by the frontier molecular energy gap between the two reactants, but also by the energy needed to distort the 1,3-dipole and dipolarophile to the transition state geometry [23]. The reactivity of dipoles and dipolarophiles can be effectively tuned by the structure and the substitution pattern, allowing modulation of the reaction kinetics.

3.1 1,3-Dipoles used in Bioorthogonal Chemistry

The idea of using dipolar cycloaddition reactions on biomolecules arose from their favorable properties, which include high atom efficiency, inherent selectivity and tunable electronics. The absence of participating functional groups in nature further underscores the great potential of this type of chemistry for bioconjugation. Moreover, significant rate enhancement of 1,3-dipolar cycloadditions observed in aqueous media [24] makes this reaction a perfect candidate for the tethering of biomolecules. Among other dipoles that react in 1,3–dipolar cycloaddition, azides, nitrile oxides, nitrones, nitrile imines, nitrile ylides, sydnones and diazo compounds have been successfully used for labeling biomolecules (Fig. 3). We will now briefly

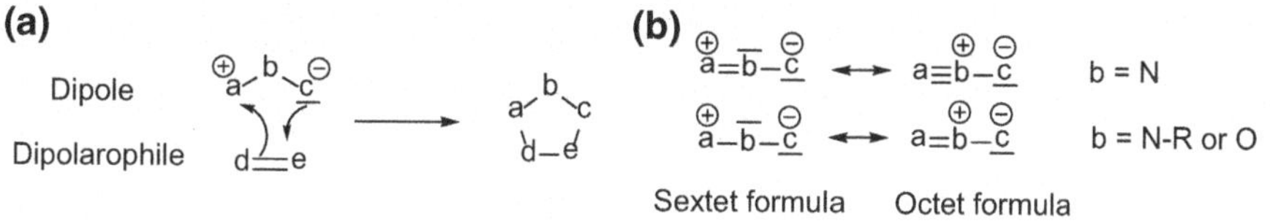

Fig. 2 **a** Schematic representation of 1,3-dipolar cycloaddition of a dipole with a dipolarophile giving rise to new five-membered ring system. **b** Resonance structures of 1,3-dipoles

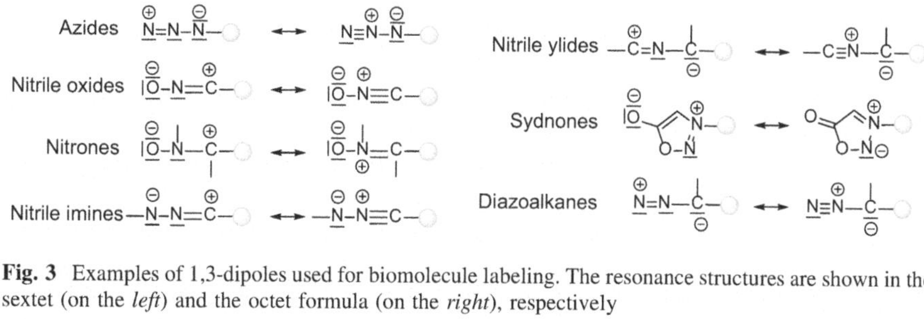

Fig. 3 Examples of 1,3-dipoles used for biomolecule labeling. The resonance structures are shown in the sextet (on the *left*) and the octet formula (on the *right*), respectively

introduce these dipoles and comment on their reactivity and applications in bioorthogonal chemistry.

3.1.1 Azides

The reaction of terminal alkynes with azides to form 1,2,3-triazoles is by far the most widely used chemistry for biomolecule labeling. This reaction, first reported in 1893 by Arthur Michael [25], and further developed by Rolf Huisgen [19], is arguably the gold standard among modern bioconjugations. The original version of the reaction requires high temperature or pressure to overcome the activation barrier. This obviously precludes its direct application to biomolecules. However, in 2002, two groups independently reported on the remarkable acceleration of the reaction by adding catalytic amounts of Cu(I) salts [26, 27]. Besides the enhanced kinetics, the Cu-catalyzed version of the reaction exclusively provides the 1,4-substituted regioisomer, in contrast to the formation of the 1,4- and 1,5-regioisomeric mixture produced by the uncatalyzed reaction (Fig. 4).

Recent studies on the reaction mechanism provided the experimental evidence for a dinuclear Cu intermediate complex involved within the reaction steps [28]. Although the Huisgen's original reaction mechanism is considered to be concerted, where the old bonds break at the same time as the new bonds form, the Cu-catalyzed azide-alkyne cycloaddition (CuAAC) is a stepwise process. Ever since the initial report, this reaction has found extensive applications in biotechnology [29–31], material sciences [32–34] and drug discovery [35, 36].

The major drawback of this methodology stems from its barriered application to biological systems as a result of the inherent toxicity of Cu(I) salts. Cu(I) promotes the generation of reactive oxygen species (ROS) that can affect the functional and structural integrity of biomolecules [37]. One commonly employed strategy to overcome this limitation is to use Cu(I)-stabilizing ligands (Fig. 5).

There is a variety of Cu(I)-stabilizing ligands currently available that can shield the active Cu(I) species and prevent the damaging effects of the catalyst to biomolecules. In addition, these ligands were found to further increase the rate of the reaction [38, 39].

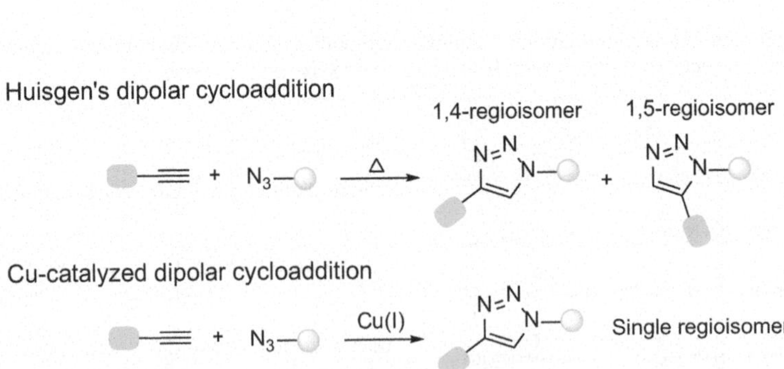

Fig. 4 The formation of regioisomers in Huisgen's original 1,3-dipolar cycloaddition vs. the Cu(I)-catalyzed version of the reaction

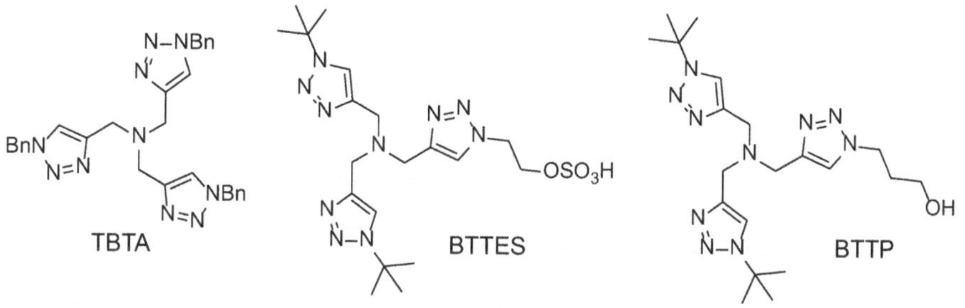

Fig. 5 Examples of Cu(I)-stabilizing ligands used in CuAAC

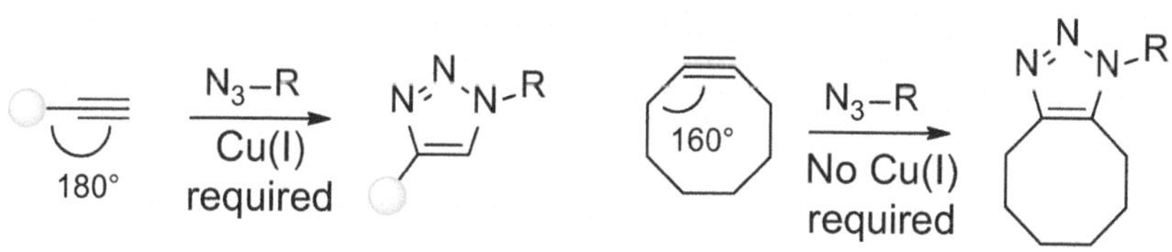

Fig. 6 Depiction showing the differences in bond angles between linear alkyne and strained cyclooctyne used in SPAAC

Another elegant strategy to improve the biocompatibility of the azide–alkyne cycloaddition, developed by the Bertozzi group [40], uses the ring strain of various cyclooctyne derivatives in lieu of toxic Cu-catalysis to boost the reaction (Fig. 6). This strain-promoted azide–alkyne cycloaddition reaction (SPAAC) was successfully applied even in living animals, and belongs to the most commonly employed strategies for bioconjugation [41, 42].

3.1.2 Nitrile Oxides and Nitrones

Further attempts to increase the kinetics and biocompatibility of 1,3-dipolar cycloadditions led organic chemists to explore alternative dipoles that react with multiple-bond reaction partners. Nitrile oxides are highly reactive dipoles that can react with various alkenes and alkynes to provide isoxazolines and isoxazoles, respectively. In the absence of a suitable reacting partner, nitrile oxides tend to dimerize to form furoxane derivatives, or can alternatively act as electrophiles. However, when generated in situ from suitable precursors such as hydroximoyl chlorides or by mild oxidation directly from oximes, nitrile oxides were successfully applied to the labeling of nucleic acids [43], peptides [44] and carbohydrates [45] (Fig. 7).

Strain-promoted alkyne–nitrone cycloaddition (SPANC) is another example of a 1,3-dipolar cycloaddition that has been applied for biomolecule labeling. The reactivity of nitrones with strained alkynes can be finely tuned by varying the substitution pattern [46–48]. Various cyclic nitrones were found to react with superior kinetics when compared to their acyclic counterparts and were successfully

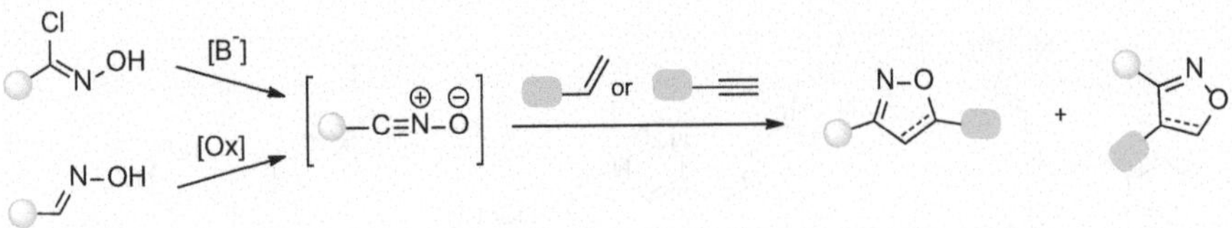

Fig. 7 In situ generated nitrile oxides from hydroximoyl chloride or oxime derivatives can react with olefins to form the corresponding regioisomeric cycloadducts

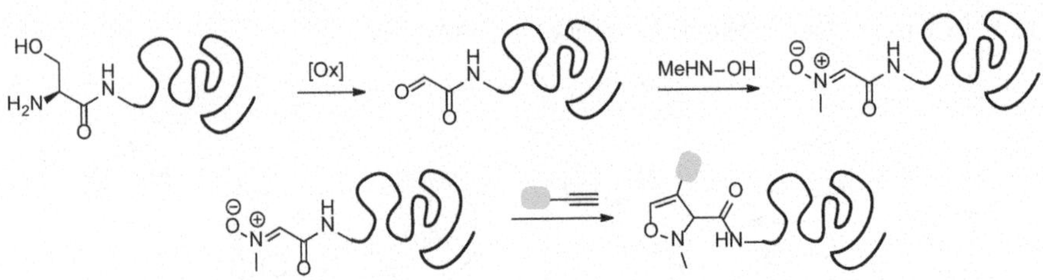

Fig. 8 N-terminal protein labeling using alkyne–nitrone 1,3-dipolar cycloaddition

applied to cell surface labeling of cancer cells [49]. An elegant strategy developed for selective protein labeling by SPANC is based on the oxidation of an N-terminal serine residue followed by reaction of the generated aldehyde with *N*-methylhydroxylamine to form the nitrone in situ. This reactive moiety subsequently reacts with a modified strained alkyne to form the desired *N*-methyl isoxazoline product on the protein (Fig. 8) [50]. A similar strategy was recently employed for N-terminal dual functionalization of peptides and proteins [51].

3.1.3 Nitrile Imines and Nitrile Ylides

The nitrile imine dipole can be generated in two ways: under basic conditions from the respective hydrazonoyl halides or from tetrazole precursors employing heat or photolysis (Fig. 9). The kinetics of the nitrile imine–alkene reaction to form pyrazoline cycloadduct strongly depends on pH and the concentration of chloride anions in the medium [52]. The discovery that photolysis of diaryl tetrazoles to form reactive nitrile imines proceeds efficiently by using only mild photoactivation (302 nm hand-held UV lamp) [53] led to successful application of this methodology to protein labeling [54]. Similarly, nitrile imines generated by mild dehydrohalogenation in aqueous buffer was used for efficient labeling of human polymerase protein [55]. An especially advantageous feature of this particular type of chemistry is the direct formation of fluorescent pyrazoline products [56]. This enables visualization of the intended biological target without the need to synthesize derivatives containing an extra fluorophore moiety and generally leads to lower background since only the product of the reaction is fluorescent. This photoinducible fluorogenic click chemistry is especially well suited for applications where careful spatiotemporal control of the studied process is required [57].

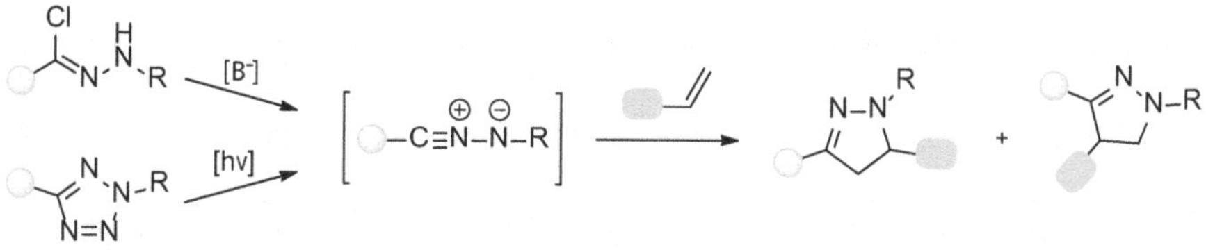

Fig. 9 Nitrile imines can be generated from hydrazonoyl chlorides or photochemically from tetrazoles. Subsequent reaction with alkenes yields fluorescent pyrazolines

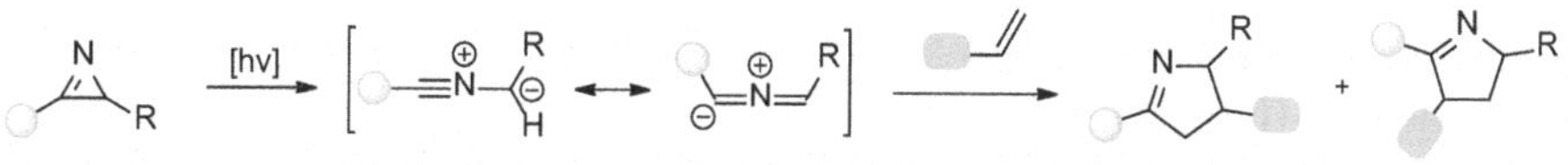

Fig. 10 Nitrile ylides can be generated by irradiation of 2*H*-azirines and react with electron deficient alkenes to give regioisomeric pyrroline products

The alluring possibility to form reactive 1,3-dipoles from suitable precursors in situ using light was further extended to nitrile ylides. These dipoles can be generated photochemically from the corresponding 2*H*-azirines (Fig. 10) [58]. This methodology was employed by Lin and coworkers to modify an azirine-containing protein with electron-deficient fumarate derivatives [59].

3.1.4 Sydnones and Diazoalkanes

Sydnones are another example of 1,3-dipoles that participate in dipolar cycloadditions with alkenes [60] and alkynes [61]. This reaction proceeds efficiently only when performed under thermal conditions. An elegant study from 2013 where high-throughput screening was employed for the discovery of new potential bioorthogonal reactions demonstrated that *N*-aryl sydnones can efficiently react with alkynes under mild conditions in the presence of Cu(I)-salts [62]. Based on the precedent that conformational ring strain in alkyne derivatives can accelerate various dipolar cycloadditions, Chin and coworkers demonstrated that sydnone cycloaddition can be used for site-specific fluorescent labeling of bicyclononyne-containing proteins (Fig. 11a) [63]. A complementary study on sydnones bearing a chlorine substituent at position 4 has shown their favorable kinetic properties. These chlorosydnone

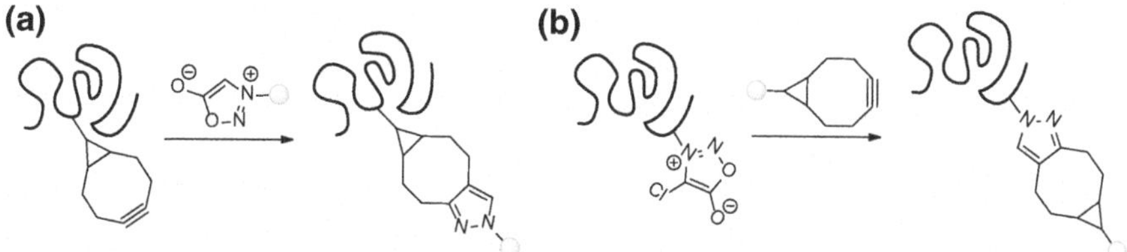

Fig. 11 **a** Bicyclononyne-containing proteins can be labeled with sydnone derivatives and vice versa, **b** Sydnone-containing proteins can be labeled with bicyclononyne derivatives

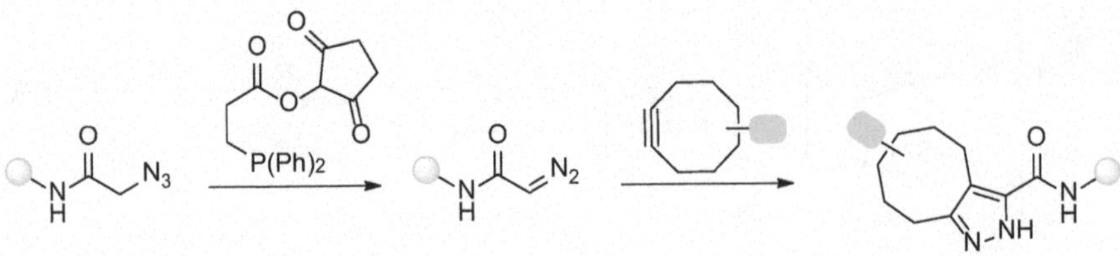

Fig. 12 Diazoalkanes generated from the corresponding azides can react with strained alkynes in 1,3-dipolar cycloaddition

derivatives attached to proteins were shown to react with fluorophore-equipped bicyclononyne (Fig. 11b) [64].

The development of a mild and efficient method that allows for the transformation of an azide into a diazo compound by using activated phosphinoesters [65] was an important step toward applying diazoalkanes for bioconjugation (Fig. 12). Systematic theoretical and experimental study on the reactivity of diazo compounds led to the identification of derivatives suitable for biomolecule labeling experiments [66]. This type of chemistry was successfully employed for cellular trafficking of diazo sugars in mammalian cells [67]. A combination of strained dibenzocyclooctyne derivative containing a cyclopropenone moiety with various diazo compounds leads to fluorescent aromatic 1*H*-pyrazoles after tautomerization. This methodology was recently used for fluorogenic protein labeling [68].

1,3-dipolar cycloadditions epitomize the rebirth of a very old reaction not initially intended to be compatible with biological systems, but with slight tweaking of the reaction conditions, it can provide tremendous potential to be just that. Since the seminal work of Rolf Huisgen over 50 years ago, organic chemists and chemical biologists all over the globe have applied and/or improved this chemistry to make 1,3-dipolar cycloadditions a cornerstone in bioorthogonal chemistry. The outlined examples of recently developed methodology clearly demonstrate the elegance and relevance of the chemistry to biomolecule labeling. Although great progress in bioorthogonal reactions was achieved in the last 15 years, the huge diversity, modularity and versatility of 1,3-dipolar cycloadditions will surely enable further improvements to this type of chemistry and allow for the full development of its potential for application to biomolecules.

4 Diels–Alder Reactions in Bioorthogonal Chemistry

Nearly 90 years ago, in 1928, Professor Otto Paul Hermann Diels and his student Kurt Alder were the first to properly identify the products of the reaction of cyclopentadiene with quinone [69]. The principal of this chemical transformation, known today as the Diels–Alder reaction, has found tremendous application in the total synthesis of complex natural products and is rightfully considered one of the most important carbon–carbon bond forming reactions in organic chemistry [70].

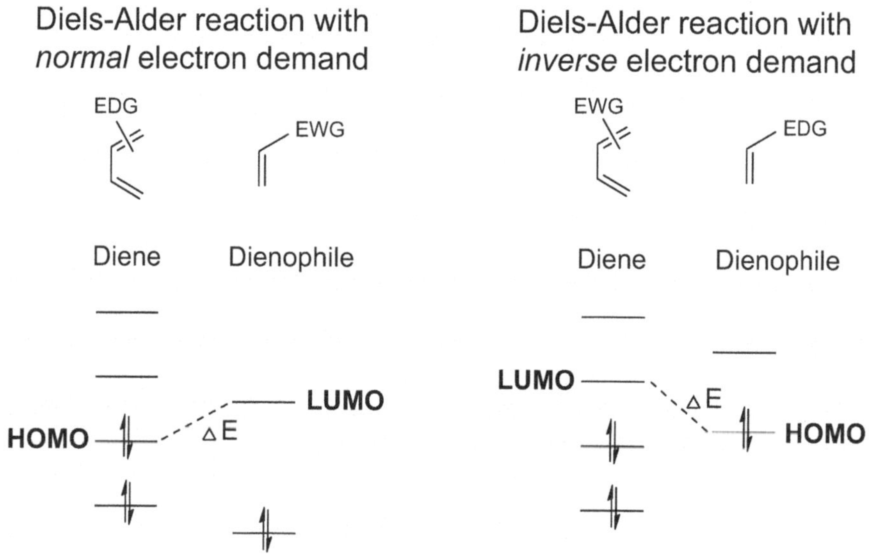

Fig. 13 Diels-Alder reactions are controlled by the energy gap between the frontier molecular orbitals

The importance of the Diels–Alder reaction was recognized by the Nobel Prize in Chemistry in 1950 [4].

The Diels–Alder (DA) reaction is a pericyclic [4 + 2]-cycloaddition reaction where a 4π electron system (a diene) reacts with a 2π electron system (a dienophile), yielding a new six-membered ring product. The reaction is stereospecific, where the stereochemistry of the starting compounds is preserved in the products. In a DA reaction with 'normal' electron demand, the dienophile typically bears an electron withdrawing substituent, while the diene is electron-rich. The case of the reverse situation, where an electron-poor diene reacts with an electron-rich dienophile, is known as the Diels–Alder reaction with inverse electron demand.

The DA reaction is controlled by the energy gap between the highest occupied molecular orbital (HOMO) and lowest unoccupied molecular orbital (LUMO) of the reactants (Fig. 13). The rate of the reaction can be significantly enhanced by using Lewis acid catalysis. An attractive feature of the DA reaction to biomolecule labeling is the observed rate enhancement in polar solvents, including water [71, 72]. The DA reaction fulfills the criteria of click reaction, and therefore plays a key role in modern bioorthogonal chemistry.

4.1 Diels–Alder Reactions with 'Normal' Electron Demand

The most commonly employed dienophiles in DA reactions with 'normal' electron demand are maleimide derivatives. These electron-poor alkenes react smoothly with various conjugated dienes to form the respective bicyclic cycloadducts under mild conditions (Fig. 14a). One drawback of maleimides that limits their widespread use in bioconjugation is their ability to react with naturally occurring nucleophiles,

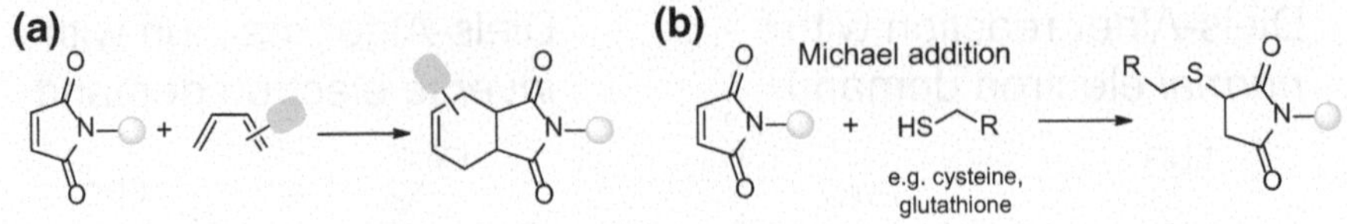

Fig. 14 **a** Maleimides participate in DA reactions with diene-containing biomolecules. **b** maleimides also react with thiol-containing biomolecules in Michael type addition

especially with thiols in a Michael-type reaction (Fig. 14b). However, this chemistry can be used for selective labeling of cysteine-containing proteins, and remains among the most commonly employed protein labeling reactions. [73]

Despite some limitations, maleimides were successfully applied in the field many times, including in the labeling of diene-containing nucleic acids [74, 75], for labeling and immobilizing proteins [76] and for the preparation of peptide–oligonucleotide conjugates [77].

In 2013, Lei's group described an elegant methodology that utilizes the in situ formation of o-quinolinone quinone methide from suitable precursor under physiological conditions. The in situ-generated methide reacts with vinyl thioethers as electron rich dienophile in hetero-Diels–Alder reaction. This chemistry was successfully used for live cell imaging [78].

The reversibility of the DA reaction, known as the retro Diels–Alder reaction, can hamper the utilization of this chemistry for bioconjugation when the formation of thermally stable products is absolutely necessary. This limitation can be conveniently overcome by the use of dienes that form stable cycloadducts during the reaction. One such example is the inverse electron-demand Diels–Alder reaction of heterodienes with strained alkenes and alkynes.

4.2 Inverse Electron Demand Diels–Alder Reactions

The reaction of an electron poor diene with an electron rich dienophile is referred to as an inverse electron demand Diels–Alder (iEDDA) reaction. The first experimental evidence that the 'normal' electron polarity of the diene/dienophile pair can be reversed dates back to 1959 [79]. Although a number of azadienes can participate in iEDDA reaction [80], the highly electron deficient nature of 1,2,4,5-tetrazines made these dienes versatile building blocks in bioconjugation applications.

In 2008, two groups independently reported on the reactivity of 1,2,4,5-tetrazines with strained olefins including *trans*-cyclooctene and norbornene [81, 82]. In the course of the reaction, the initially formed highly strained bicyclic adduct rapidly converts to the 4,5-dihydropyridazine. The final prototropic isomerization leads to the corresponding 1,4-dihydro product that can be further oxidized (external oxidants are usually required) to the fully aromatic pyridazine. A molecule of nitrogen is produced by the reaction as the only by-product (Fig. 15).

Besides *trans*-cyclooctenes and norbornenes, 1,2,4,5-tetrazines also react with other dienophiles that include: cyclooctynes [83], cyclopropenes [84, 85] and *N*-

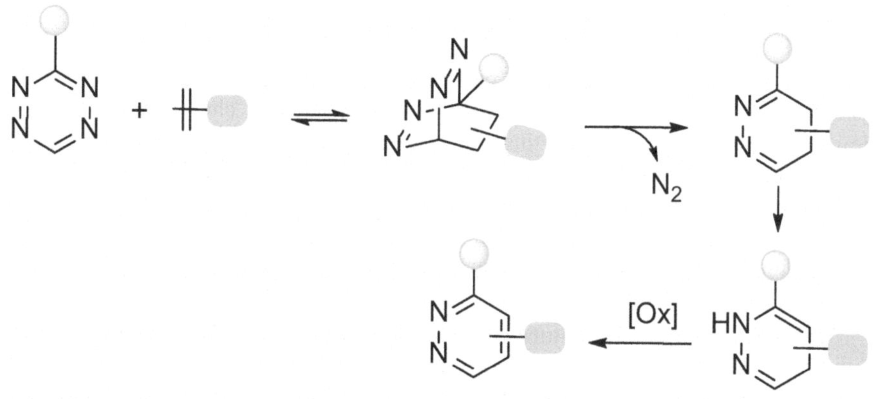

Fig. 15 Reaction scheme of the iEDDA of 1,2,4,5-tetrazines

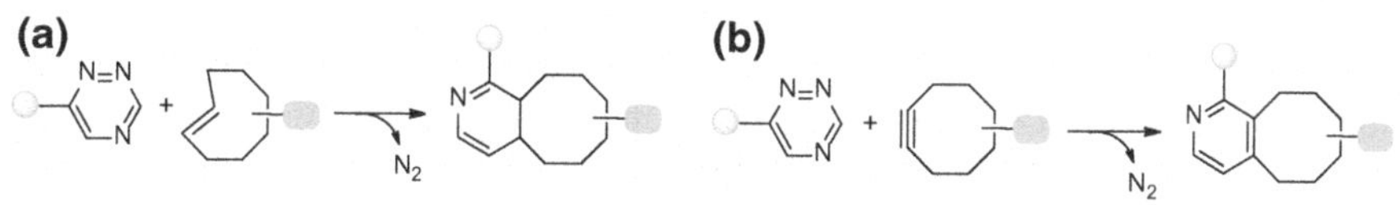

Fig. 16 Reaction of 1,2,4-triazines with **a** *trans*-cyclooctene and **b** with a strained alkyne

acetylazetines [86]. Due to the extrusion of nitrogen, the irreversible nature of this chemical transformation enables the production of highly stable bioconjugates in a straightforward manner. The extraordinarily fast reaction kinetics of iEDDA reactions (in fact, the fastest bioorthogonal reaction reported to date) enabled utilization of this chemistry for efficient labeling of biomolecules that include proteins, nucleic acids and glycans [87].

Concerns about the stability of 1,2,4,5-tetrazines under physiological conditions [88] and their problematic synthetic accessibility led researchers to look for alternative azadienes that participate in reactions with activated dienophiles. Based on computational analysis of the reactivity of 1,2,3- and 1,2,4- triazines with strained olefins led to the assumption that 1,2,4- triazines should participate in the iEDDA reaction rather than 1,2,3- triazines. Indeed, a series of mono-aryl 1,2,4-triazines were found to react with *trans*-cyclooctene to afford the corresponding dihydropyridine products (Fig. 16a). This methodology was successfully used for labeling triazine-containing proteins [89]. Although 1,2,4-triazines were found to be reactive toward highly strained *trans*-cyclooctene, they do not react with other strained olefins such as norbornene or cyclopropene. Since these strained alkenes are known to undergo efficient cycloadditions with 1,2,4,5-tetrazines, this augments the likelihood of sequential dual labeling of biomolecules. However, the full potential of this methodology in this direction remains unexplored. A recent complementary study showed that besides *trans*-cyclooctene, 1,2,4-triazines also react with strained alkynes (Fig. 16b) [90].

5 Conclusions and Outlook

The vast potential of 1,3-dipolar cycloadditions and Diels–Alder reactions in synthetic organic chemistry has been exploited by organic chemists for decades. However, it was only recently that the power of these chemical transformations attracted chemical biologists who recognized the beauty and immense possibilities of the chemistry for applications on biomolecules. The inherent features of these reactions, which include fast kinetics, high yields and excellent biocompatibility, make them perfectly suited to tag and study biomolecules in their native environment.

Modern bioorthogonal reactions equip synthetic chemists and chemical biologists with unique chemical tools that tremendously extend our ability to understand and manipulate biomolecules, the key players in physiological processes. Despite the great progress that has been achieved in bioconjugation reactions there are still challenges that remain to be addressed. The synthetic accessibility of many derivatives used for bioconjugations is still far from being optimal. Extensive and non-trivial synthetic steps are often required to prepare reasonable amounts of the required compounds. On the other hand, the increasing popularity of bioorthogonal reactions also invoke the interest of commercial suppliers, and, indeed, many useful reagents are now commercially available. The high reactivity of some derivatives used for bioconjugations (e.g. *trans*-cyclooctenes, some tetrazines or strained cyclooctynes) is often inextricably linked to their reduced stability. Further optimizations of the structures are needed to preserve their high reactivity and to improve stability.

The recent advances in bioorthogonal chemistry provide us with a variety of options, where the right choice of the bioconjugation reaction for a particular application plays an important role [91]. A welcoming result that has sprung from the diversity of available bioorthogonal transformations is the possibility for mutual combinations. In fact, many of the discovered ligation methods are orthogonal to each other [92, 93]. This provides a fantastic opportunity for installing multiple tags on the same target biomolecule. New discoveries toward this direction will certainly extend the application potential of bioorthogonal ligations and will enable us to answer even more complex biological questions. We believe that the appealing opportunities that these novel methodologies provide will attract the interest of researchers far beyond the current experts in the field.

Another important requirement for successful application of bioorthogonal reactions to answer biological questions relies on our ability to introduce one of the reactive unnatural functional groups to the intended biomolecule. By moving to in vivo applications, it will be necessary to develop new methodologies that will enable precise control over the location of the chemical handle within the studied biomolecule. This is especially challenging in whole organisms. In this regard, recent advances in genetic code expansion techniques may open the door toward this goal [94, 95]. Alternatively, the design of new caged metabolic precursors that can pass the biosynthetic machinery of an organism and can be subsequently incorporated into particular tissues holds a great promise [96]. Another elegant strategy may come from synthetic biology. With the huge amount of progress that has been achieved in recent years, we envision that this particular field may provide new possibilities and ideas in the field. An organism able to synthesize non-natural building blocks on its own and also able to directly incorporate these chemical groups to specific parts of its own biomolecules would be a true revolution in this regard. We believe that these and other new methodologies will appear and enable in the near future the design and production of well-defined biological structures containing chemical functional groups suitable for bioconjugations that cannot be found in nature.

The following book chapters aim to recapitulate the recent advances in bioorthogonal reactions, with a focus on 1,3-dipolar cycloadditions and Diels–Alder reactions that were particularly influential in the field. Besides cycloaddition reactions, there are other types of chemical reactions that are selective and orthogonal to natural systems and were successfully used for various applications in chemical biology. We refer readers with broader interest on the topic to excellent reviews, published recently [97–100].

References

1. Kolb HC, Finn MG, Sharpless KB (2001) Click chemistry: diverse chemical function from a few good reactions. Angew Chem Int Ed 40(11):2004–2021. doi:10.1002/1521-3773(20010601)40:11<2004::AID-ANIE2004>3.0.CO;2-5
2. Dedecker P, De Schryver FC, Hofkens J (2013) Fluorescent proteins: shine on, you crazy diamond. J Am Chem Soc 135(7):2387–2402. doi:10.1021/ja309768d
3. Nienhaus K, Nienhaus GU (2014) Fluorescent proteins for live-cell imaging with super-resolution. Chem Soc Rev 43(4):1088–1106. doi:10.1039/c3cs60171d
4. http://nobelprize.org/nobel_prizes/chemistry/web
5. Hang HC, Yu C, Kato DL, Bertozzi CR (2003) A metabolic labeling approach toward proteomic analysis of mucin-type O-linked glycosylation. P Natl Acad Sci USA 100(25):14846–14851. doi:10.1073/pnas.2335201100
6. Rideout D, Calogeropoulou T, Jaworski J, McCarthy M (1990) Synergism through direct covalent bonding between agents: a strategy for rational design of chemotherapeutic combinations. Biopolymers 29(1):247–262. doi:10.1002/bip.360290129
7. Griffin BA, Adams SR, Tsien RY (1998) Specific covalent labeling of recombinant protein molecules inside live cells. Science 281(5374):269–272. doi:10.1126/science.281.5374.269
8. Prescher JA, Bertozzi CR (2005) Chemistry in living systems. Nat Chem Biol 1(1):13–21. doi:10.1038/nchembio0605-13
9. Sletten EM, Bertozzi CR (2009) Bioorthogonal chemistry: fishing for selectivity in a sea of functionality. Angew Chem Int Ed 48(38):6974–6998. doi:10.1002/anie.200900942
10. Merkel M, Peewasan K, Arndt S, Ploschik D, Wagenknecht HA (2015) Copper-free postsynthetic labeling of nucleic acids by means of bioorthogonal reactions. ChemBioChem 16(11):1541–1553. doi:10.1002/cbic.201500199
11. Tang W, Becker ML (2014) "Click" reactions: a versatile toolbox for the synthesis of peptide-conjugates. Chem Soc Rev 43(20):7013–7039. doi:10.1039/c4cs00139g
12. Hocek M (2014) Synthesis of base-modified 2'-deoxyribonucleoside triphosphates and their use in enzymatic synthesis of modified DNA for applications in bioanalysis and chemical biology. J Org Chem 79(21):9914–9921. doi:10.1021/jo5020799
13. Davis L, Chin JW (2012) Designer proteins: applications of genetic code expansion in cell biology. Nat Rev Mol Cell Bio 13(3):168–182. doi:10.1038/nrm3286

14. Dumas A, Lercher L, Spicer CD, Davis BG (2015) Designing logical codon reassignment—expanding the chemistry in biology. Chem Sci 6(1):50–69. doi:10.1039/c4sc01534g
15. Du J, Meledeo MA, Wang ZY, Khanna HS, Paruchuri VDP, Yarema KJ (2009) Metabolic glycoengineering: sialic acid and beyond. Glycobiology 19(12):1382–1401. doi:10.1093/glycob/cwp115
16. Jao CY, Roth M, Welti R, Salic A (2009) Metabolic labeling and direct imaging of choline phospholipids in vivo. P Natl Acad Sci USA 106(36):15332–15337. doi:10.1073/pnas.0907864106
17. Rieder U, Luedtke NW (2014) Alkene-Tetrazine ligation for imaging cellular DNA. Angew Chem Int Ed 53(35):9168–9172. doi:10.1002/anie.201403580
18. Shieh P, Siegrist MS, Cullen AJ, Bertozzi CR (2014) Imaging bacterial peptidoglycan with near-infrared fluorogenic azide probes. P Natl Acad Sci USA 111(15):5456–5461. doi:10.1073/pnas.1322727111
19. Huisgen R (1963) 1,3-dipolar cycloadditions. past and future. Angew Chem Int Ed 2(10):565–598. doi:10.1002/anie.196305651
20. Najera C, Sansano JM (2009) 1,3-Dipolar cycloadditions: applications to the synthesis of antiviral agents. Org Biomol Chem 7(22):4567–4581. doi:10.1039/b913066g
21. Padwa A, Pearson WH (2002) Synthetic applications of 1,3-dipolar cycloaddition chemistry toward heterocycles and natural products. Chem Heterocylcl Comp 59: John Wiley & Sons, Inc
22. Pearson WH (2002) Alkaloid synthesis via [3 + 2] cycloadditions. Pure Appl Chem 74(8):1339–1347. doi:10.1351/pac200274081339
23. Ess DH, Houk KN (2008) Theory of 1,3-dipolar cycloadditions: distortion/interaction and frontier molecular orbital models. J Am Chem Soc 130(31):10187–10198. doi:10.1021/ja800009z
24. Butler RN, Cunningham WJ, Coyne AG, Burke LA (2004) The influence of water on the rates of 1,3-dipolar cycloaddition reactions: trigger points for exponential rate increases in water-organic solvent mixtures. Water-super versus water-normal dipolarophiles. J Am Chem Soc 126(38):11923–11929. doi:10.1021/ja040119y
25. Michael A (1893) Ueber die Einwirkung von Diazobenzolimid auf Acetylendicarbonsäuremethylester. J Prakt Chem 48:94–95. doi:10.1002/prac.18930480114
26. Rostovtsev VV, Green LG, Fokin VV, Sharpless KB (2002) A stepwise Huisgen cycloaddition process: copper(I)-catalyzed regioselective "ligation" of azides and terminal alkynes. Angew Chem Int Ed 41(14):2596–2599. doi:10.1002/1521-3773(20020715)41:14<2596::Aid-Anie2596>3.0.Co;2-4
27. Tornoe CW, Christensen C, Meldal M (2002) Peptidotriazoles on solid phase: [1-3]-triazoles by regiospecific copper(I)-catalyzed 1,3-dipolar cycloadditions of terminal alkynes to azides. J Org Chem 67(9):3057–3064. doi:10.1021/jo011148j
28. Worrell BT, Malik JA, Fokin VV (2013) Direct evidence of a dinuclear copper intermediate in Cu(I)-catalyzed azide-alkyne cycloadditions. Science 340(6131):457–460. doi:10.1126/science.1229506
29. Lahann J (2009) Click chemistry for biotechnology and materials science. Wiley, Hoboken. doi:10.1002/9780470748862
30. El-Sagheer AH, Brown T (2010) Click chemistry with DNA. Chem Soc Rev 39(4):1388–1405. doi:10.1039/b901971p
31. Lallana E, Riguera R, Fernandez-Megia E (2011) Reliable and efficient procedures for the conjugation of biomolecules through huisgen azide-alkyne cycloadditions. Angew Chem Int Ed 50(38):8794–8804. doi:10.1002/anie.201101019
32. Hanni KD, Leigh DA (2010) The application of CuAAC 'click' chemistry to catenane and rotaxane synthesis. Chem Soc Rev 39(4):1240–1251. doi:10.1039/b901974j
33. Ganesh V, Sudhir VS, Kundu T, Chandrasekaran S (2011) 10 years of click chemistry: synthesis and applications of ferrocene-derived triazoles. Chem-Asian J 6(10):2670–2694. doi:10.1002/asia.201100408
34. Johnson RP, John JV, Kim I (2013) Recent developments in polymer-block-polypeptide and protein-polymer bioconjugate hybrid materials. Eur Polym J 49(10):2925–2948. doi:10.1016/j.eurpolymj.2013.04.017
35. Tron GC, Pirali T, Billington RA, Canonico PL, Sorba G, Genazzani AA (2008) Click chemistry reactions in medicinal chemistry: applications of the 1,3-dipolar cycloaddition between azides and alkynes. Med Res Rev 28(2):278–308. doi:10.1002/med.20107
36. Thirumurugan P, Matosiuk D, Jozwiak K (2013) Click chemistry for drug development and diverse chemical-biology applications. Chem Rev 113(7):4905–4979. doi:10.1021/cr200409f

37. Kennedy DC, McKay CS, Legault MCB, Danielson DC, Blake JA, Pegoraro AF, Stolow A, Mester Z, Pezacki JP (2011) Cellular consequences of copper complexes used to catalyze bioorthogonal click reactions. J Am Chem Soc 133(44):17993–18001. doi:10.1021/ja2083027
38. Besanceney-Webler C, Jiang H, Zheng TQ, Feng L, del Amo DS, Wang W, Klivansky LM, Marlow FL, Liu Y, Wu P (2011) Increasing the efficacy of bioorthogonal click reactions for bioconjugation: a comparative study. Angew Chem Int Ed 50(35):8051–8056. doi:10.1002/anie.201101817
39. Wang W, Hong SL, Tran A, Jiang H, Triano R, Liu Y, Chen X, Wu P (2011) Sulfated ligands for the copper(I)-catalyzed azide-alkyne cycloaddition. Chem-Asian J 6(10):2796–2802. doi:10.1002/asia.201100385
40. Agard NJ, Prescher JA, Bertozzi CR (2004) A strain-promoted [3 + 2] azide-alkyne cycloaddition for covalent modification of blomolecules in living systems. J Am Chem Soc 126(46):15046–15047. doi:10.1021/ja0449981
41. Baskin JM, Prescher JA, Laughlin ST, Agard NJ, Chang PV, Miller IA, Lo A, Codelli JA, Bertozzi CR (2007) Copper-free click chemistry for dynamic in vivo imaging. P Natl Acad Sci USA 104(43):16793–16797. doi:10.1073/pnas.0707090104
42. Laughlin ST, Bertozzi CR (2009) In vivo imaging of caenorhabditis elegans Glycans. ACS Chem Biol 4(12):1068–1072. doi:10.1021/cb900254y
43. Gutsmiedl K, Wirges CT, Ehmke V, Carell T (2009) Copper-free "click" modification of DNA via nitrile oxide-norbornene 1,3-dipolar cycloaddition. Org Lett 11(11):2405–2408. doi:10.1021/ol9005322
44. Jawalekar AM, Reubsaet E, Rutjes FPJT, van Delft FL (2011) Synthesis of isoxazoles by hypervalent iodine-induced cycloaddition of nitrile oxides to alkynes. Chem Commun 47(11):3198–3200. doi:10.1039/c0cc04646a
45. Sanders BC, Friscourt F, Ledin PA, Mbua NE, Arumugam S, Guo J, Boltje TJ, Popik VV, Boons GJ (2011) Metal-free sequential [3 + 2]-dipolar cycloadditions using cyclooctynes and 1,3-dipoles of different reactivity. J Am Chem Soc 133(4):949–957. doi:10.1021/ja1081519
46. Mckay CS, Moran J, Pezacki JP (2010) Nitrones as dipoles for rapid strain-promoted 1,3-dipolar cycloadditions with cyclooctynes. Chem Commun 46(6):931–933. doi:10.1039/b921630h
47. McKay CS, Chigrinova M, Blake JA, Pezacki JP (2012) Kinetics studies of rapid strain-promoted [3 + 2]-cycloadditions of nitrones with biaryl-aza-cyclooctynone. Org Biomol Chem 10(15):3066–3070. doi:10.1039/c2ob07165g
48. MacKenzie DA, Pezacki JP (2014) Kinetics studies of rapid strain-promoted [3 + 2] cycloadditions of nitrones with bicyclo[6.1.0]nonyne. Can J Chem 92(4):337–340. doi:10.1139/cjc-2013-0577
49. McKay CS, Blake JA, Cheng J, Danielson DC, Pezacki JP (2011) Strain-promoted cycloadditions of cyclic nitrones with cyclooctynes for labeling human cancer cells. Chem Commun 47(36):10040–10042. doi:10.1039/c1cc13808a
50. Ning XH, Temming RP, Dommerholt J, Guo J, Ania DB, Debets MF, Wolfert MA, Boons GJ, van Delft FL (2010) Protein modification by strain-promoted alkyne-nitrone cycloaddition. Angew Chem Int Ed 49(17):3065–3068. doi:10.1002/anie.201000408
51. Temming RP, Eggermont L, van Eldijk MB, van Hest JCM, van Delft FL (2013) N-terminal dual protein functionalization by strain-promoted alkyne-nitrone cycloaddition. Org Biomol Chem 11(17):2772–2779. doi:10.1039/c3ob00043e
52. Wang XS, Lee YJ, Liu WSR (2014) The nitrilimine-alkene cycloaddition is an ultra rapid click reaction. Chem Commun 50(24):3176–3179. doi:10.1039/c3cc48682f
53. Wang YZ, Vera CIR, Lin Q (2007) Convenient synthesis of highly functionalized pyrazolines via mild, photoactivated 1,3-dipolar cycloaddition. Org Lett 9(21):4155–4158. doi:10.1021/ol7017328
54. Song W, Wang Y, Qu J, Madden MM, Lin Q (2008) A photoinducible 1,3-dipolar cycloaddition reaction for rapid, selective modification of tetrazole-containing proteins. Angew Chem Int Ed 47(15):2832–2835. doi:10.1002/anie.200705805
55. Kaya E, Vrabel M, Deiml C, Prill S, Fluxa VS, Carell T (2012) A genetically encoded norbornene amino acid for the mild and selective modification of proteins in a copper-free click reaction. Angew Chem Int Ed 51(18):4466–4469. doi:10.1002/anie.201109252
56. Ramil CP, Lin Q (2014) Photoclick chemistry: a fluorogenic light-triggered in vivo ligation reaction. Curr Opin Chem Biol 21:89–95. doi:10.1016/j.cbpa.2014.05.024
57. Lim RKV, Lin Q (2011) Photoinducible bioorthogonal chemistry: a spatiotemporally controllable tool to visualize and perturb proteins in live cells. Acc Chem Res 44(9):828–839. doi:10.1021/ar200021p

58. Padwa A, Smolanof J (1971) Photocycloaddition of arylazirenes with electron-deficient olefins. J Am Chem Soc 93(2):548–550. doi:10.1021/ja00731a056
59. Lim RKV, Lin Q (2010) Azirine ligation: fast and selective protein conjugation via photoinduced azirine-alkene cycloaddition. Chem Commun 46(42):7993–7995. doi:10.1039/c0cc02863k
60. Huisgen R, Gotthardt H, Grashey R (1962) Reactions of sydnones with alkenes. Angew Chem Int Ed 1(1):49. doi:10.1002/anie.196200491
61. Huisgen R, Grashey R, Gotthardt H, Schmidt R (1962) 1,3-dipolar additions of sydnones to alkynes. a new route into the pyrazole series. Angew Chem Int Ed 1(1):48–49. doi:10.1002/anie.196200482
62. Kolodych S, Rasolofonjatovo E, Chaumontet M, Nevers MC, Creminon C, Taran F (2013) Discovery of chemoselective and biocompatible reactions using a high-throughput immunoassay screening. Angew Chem Int Ed 52(46):12056–12060. doi:10.1002/anie.201305645
63. Wallace S, Chin JW (2014) Strain-promoted sydnone bicyclo-[6.1.0]-nonyne cycloaddition. Chem Sci 5(5):1742–1744. doi:10.1039/c3sc53332h
64. Plougastel L, Koniev O, Specklin S, Decuypere E, Creminon C, Buisson DA, Wagner A, Kolodych S, Taran F (2014) 4-Halogeno-sydnones for fast strain promoted cycloaddition with bicyclo-[6.1.0]-nonyne. Chem Commun 50(66):9376–9378. doi:10.1039/c4cc03816a
65. Myers EL, Raines RT (2009) A phosphine-mediated conversion of azides into diazo compounds. Angew Chem Int Ed 48(13):2359–2363. doi:10.1002/anie.200804689
66. McGrath NA, Raines RT (2012) Diazo compounds as highly tunable reactants in 1,3-dipolar cycloaddition reactions with cycloalkynes. Chem Sci 3(11):3237–3240. doi:10.1039/c2sc20806g
67. Andersen KA, Aronoff MR, McGrath NA, Raines RT (2015) Diazo groups endure metabolism and enable chemoselectivity in cellulo. J Am Chem Soc 137(7):2412–2415. doi:10.1021/ja5095815
68. Friscourt F, Fahrni CJ, Boons G-J (2015) Fluorogenic strain-promoted alkyne-diazo cycloadditions. Chem A Eur J 21(40):13996–14001. doi:10.1002/chem.201502242
69. Diels O, Alder K (1928) Synthesen in der hydroaromatischen Reihe. Justus Liebigs Annalen der Chemie 460(1):98–122. doi:10.1002/jlac.19284600106
70. Nicolaou KC, Snyder SA, Montagnon T, Vassilikogiannakis G (2002) The Diels-Alder reaction in total synthesis. Angew Chem Int Ed 41(10):1668–1698. doi:10.1002/1521-3773(20020517)41:10<1668::Aid-Anie1668>3.0.Co;2-Z
71. Rideout DC, Breslow R (1980) Hydrophobic acceleration of diels-alder reactions. J Am Chem Soc 102(26):7816–7817. doi:10.1021/ja00546a048
72. Otto S, Blokzijl W, Engberts JBFN (1994) Diels-alder reactions in water—effects of hydrophobicity and hydrogen-bonding. J Org Chem 59(18):5372–5376. doi:10.1021/jo00097a045
73. Chalker JM, Bernardes GJL, Lin YA, Davis BG (2009) Chemical modification of proteins at cysteine: opportunities in chemistry and biology. Chem-Asian J 4(5):630–640. doi:10.1002/asia.200800427
74. Seelig B, Jaschke A (1997) Site-specific modification of enzymatically synthesized RNA: transcription initiation and Diels-Alder reaction. Tetrahedron Lett 38(44):7729–7732. doi:10.1016/S0040-4039(97)10151-4
75. Hill KW, Taunton-Rigby J, Carter JD, Kropp E, Vagle K, Pieken W, McGee DPC, Husar GM, Leuck M, Anziano DJ, Sebesta DP (2001) Diels-Alder bioconjugation of diene-modified oligonucleotides. J Org Chem 66(16):5352–5358. doi:10.1021/jo0100190
76. de Araujo AD, Palomo JM, Cramer J, Kohn M, Schroder H, Wacker R, Niemeyer C, Alexandrov K, Waldmann H (2006) Diels-Alder ligation and surface immobilization of proteins. Angew Chem Int Ed 45(2):296–301. doi:10.1002/anie.200502266
77. Marchan V, Ortega S, Pulido D, Pedroso E, Grandas A (2006) Diels-Alder cycloadditions in water for the straightforward preparation of peptide-oligonucleotide conjugates. Nucleic Acids Res 34(3):e24. doi:10.1093/nar/gnj020
78. Li Q, Dong T, Liu X, Lei X (2013) A bioorthogonal ligation enabled by click cycloaddition of o-Quinolinone Quinone methide and vinyl thioether. J Am Chem Soc 135(13):4996–4999. doi:10.1021/ja401989p
79. Carboni RA, Lindsey RV (1959) Reactions of tetrazines with unsaturated compounds. A new synthesis of pyridazines. J Am Chem Soc 81(16):4342–4346. doi:10.1021/ja01525a060
80. Boger DL (1986) Diels-Alder reactions of heterocyclic aza dienes, Scope and applications. Chem Rev 86(5):781–793. doi:10.1021/cr00075a004
81. Blackman ML, Royzen M, Fox JM (2008) Tetrazine ligation: fast bioconjugation based on inverse-electron-demand Diels-Alder reactivity. J Am Chem Soc 130(41):13518–13519. doi:10.1021/ja8053805

82. Devaraj NK, Weissleder R, Hilderbrand SA (2008) Tetrazine-based cycloadditions: application to pretargeted live cell imaging. Bioconjugate Chem 19(12):2297–2299. doi:10.1021/bc8004446
83. Chen WX, Wang DZ, Dai CF, Hamelberg D, Wang BH (2012) Clicking 1,2,4,5-tetrazine and cyclooctynes with tunable reaction rates. Chem Commun 48(12):1736–1738. doi:10.1039/c2cc16716f
84. Patterson DM, Nazarova LA, Xie B, Kamber DN, Prescher JA (2012) Functionalized cyclopropenes as bioorthogonal chemical reporters. J Am Chem Soc 134(45):18638–18643. doi:10.1021/ja3060436
85. Yang J, Seckute J, Cole CM, Devaraj NK (2012) Live-cell imaging of cyclopropene tags with fluorogenic tetrazine cycloadditions. Angew Chem Int Ed 51(30):7476–7479. doi:10.1002/anie.201202122
86. Engelsma SB, Willems LI, van Paaschen CE, van Kasteren SI, van der Marel GA, Overkleeft HS, Filippov DV (2014) Acylazetine as a dienophile in bioorthogonal inverse electron-demand diels-alder ligation. Org Lett 16(10):2744–2747. doi:10.1021/ol501049c
87. Knall AC, Slugovc C (2013) Inverse electron demand Diels-Alder (iEDDA)-initiated conjugation: a (high) potential click chemistry scheme. Chem Soc Rev 42(12):5131–5142. doi:10.1039/c3cs60049a
88. Kämpchen T, Massa W, Overheu W, Schmidt R, Seitz G (1982) Zur Kenntnis von reaktionen des 1,2,4,5-tetrazin-3,6-dicarbonsäure-dimethylesters mit nucleophilen. Chem Ber 115(2):683–694. doi:10.1002/cber.19821150228
89. Kamber DN, Liang Y, Blizzard RJ, Liu F, Mehl RA, Houk KN, Prescher JA (2015) 1,2,4-triazines are versatile bioorthogonal reagents. J Am Chem Soc 137(26):8388–8391. doi:10.1021/jacs.5b05100
90. Horner KA, Valette NM, Webb ME (2015) Strain-promoted reaction of 1,2,4-triazines with bicyclononynes. istry. Chem A Eur J 21(41):14376–14381. doi:10.1002/chem.201502397
91. Patterson DM, Nazarova LA, Prescher JA (2014) Finding the right (Bioorthogonal) chemistry. ACS Chem Biol 9(3):592–605. doi:10.1021/cb400828a
92. Beal DM, Jones LH (2012) Molecular scaffolds using multiple orthogonal conjugations: applications in chemical biology and drug discovery. Angew Chem Int Ed 51(26):6320–6326. doi:10.1002/anie.201200002
93. Tunca U (2014) Orthogonal multiple click reactions in synthetic polymer chemistry. J Polym Sci Pol Chem 52(22):3147–3165. doi:10.1002/pola.27379
94. Greiss S, Chin JW (2011) Expanding the genetic code of an animal. J Am Chem Soc 133(36):14196–14199. doi:10.1021/ja2054034
95. Bianco A, Townsley FM, Greiss S, Lang K, Chin JW (2012) Expanding the genetic code of Drosophila melanogaster. Nat Chem Biol 8(9):748–750. doi:10.1038/Nchembio.1043
96. Chang PV, Dube DH, Sletten EM, Bertozzi CR (2010) A Strategy for the selective imaging of glycans using caged metabolic precursors. J Am Chem Soc 132(28):9516–9518. doi:10.1021/ja101080y
97. King M, Wagner A (2014) Developments in the field of bioorthogonal bond forming reactions-past and present trends. Bioconjugate Chem 25(5):825–839. doi:10.1021/bc500028d
98. McKay CS, Finn MG (2014) Click chemistry in complex mixtures: bioorthogonal bioconjugation. Chem Biol 21(9):1075–1101. doi:10.1016/j.chembiol.2014.09.002
99. Spicer CD, Davis BG (2014) Selective chemical protein modification. Nat Commun. doi:10.1038/ncomms5740
100. Boutureira O, Bernardes GJL (2015) Advances in chemical protein modification. Chem Rev 115(5):2174–2195. doi:10.1021/cr500399p

Top Curr Chem (Z) (2016) 374:2
DOI 10.1007/s41061-015-0001-3

REVIEW

Transition-Metal-Catalyzed Bioorthogonal Cycloaddition Reactions

Maiyun Yang[1] · Yi Yang[1] · Peng R. Chen[1,2]

Received: 26 October 2015 / Accepted: 11 November 2015 / Published online: 14 December 2015

Abstract In recent years, bioorthogonal reactions have emerged as a powerful toolbox for specific labeling and visualization of biomolecules, even within the highly complex and fragile living systems. Among them, copper(I)-catalyzed azide–alkyne cycloaddition (CuAAC) reaction is one of the most widely studied and used biocompatible reactions. The cytotoxicity of Cu(I) ions has been greatly reduced due to the use of Cu(I) ligands, which enabled the CuAAC reaction to proceed on the cell surface, as well as within an intracellular environment. Meanwhile, other transition metals such as ruthenium, rhodium and silver are now under development as alternative sources for catalyzing bioorthogonal cycloadditions. In this review, we summarize the development of CuAAC reaction as a prominent bioorthogonal reaction, discuss various ligands used in reducing Cu(I) toxicity while promoting the reaction rate, and illustrate some of its important biological applications. The development of additional transition metals in catalyzing cycloaddition reactions will also be briefly introduced.

Keywords Bioorthogonal reaction · Transition metal · Cycloaddition reaction · CuAAC reaction · Unnatural amino acids

---

Maiyun Yang and Yi Yang have contributed equally to this work.

✉ Peng R. Chen
pengchen@pku.edu.cn

Maiyun Yang
maiyun123@163.com

1 Synthetic and Functional Biomolecule Center, College of Chemistry and Molecular Engineering, Peking University, Beijing 100871, China

2 Peking-Tsinghua Center for Life Sciences, Beijing 100871, China

1 Introduction

Cycloaddition reactions occupy an important position in the arsenal of both synthetic chemistry and bioorthogonal chemistry. In the field of bioorthogonal reactions, cycloadditions can be promoted by transition metals (e.g. copper(I)-catalyzed azide–alkyne cycloaddition [CuAAC] reaction [1–5]), light (e.g. photo-induced tetrazole-alkene cycloaddition [6, 7]), or by strain activation (e.g. strain-promoted azide–alkyne cycloaddition [SPAAC] [8], inverse electron-demand Diels–Alder [IEDDA] reaction [9, 10]). These cycloaddition reactions have been widely adopted by chemical biologists to study biological questions in living systems [11–13]; however, there are still many concerns to be addressed. First, the transition metals and ultraviolet light can cause cytotoxicity to live cells and tissues. Second, there are concerns regarding the bioorthogonality of some of these functional groups in live cells. For example, the strained cyclooctyne may undergo side reactions with cellular nucleophiles [14]. Third, the synthesis procedures for some of the functional group-containing reagents are complicated, such as the tetrazine- and trans-cyclooctene-containing compounds, which may prevent their broad utility at the current stage.

Compared with strain- and light-triggered cycloaddition, the bioorthogonal reaction pair utilized in CuAAC is quite small in size, which enables its fast cellular uptake and small perturbation to target biomolecules. Meanwhile, the azide and alkyne functional groups are inert to the surrounding biological environment and remain mutually reactive with a high reaction rate (Table 1). These features possess unique advantages, which allow CuAAC to be widely applied in living systems to label and manipulate intact biomolecules [29–31]. Nevertheless, Cu(I) ions have significant cytotoxicity to live cells, which hindered its direct usage inside living systems. To avoid this problem, various synthetic ligands have been developed to prevent the harmful effects of Cu(I) ions. These ligand-assisted CuAAC reactions were first adopted on cell surface labeling, and then moved to label biomolecules in the intracellular space with improved biocompatibility. Meanwhile, efforts have also been put to explore additional transition metals such as ruthenium (Ru), rhodium (Rh) and silver (Ag) to substitute Cu(I) ions to promote cycloaddition. In this review, we present recent progresses in transition-metal-catalyzed bioorthogonal cycloaddition reactions, as well as their biological applications.

2 Copper(I)-Catalyzed Azide–Alkyne Cycloaddition (CuAAC)

2.1 Development of CuAAC

Azide–alkyne reaction has been known since 1893 when A. Michael reported the first synthesis of 1,2,3-triazoles from diethyl acetylenedicarboxylate and phenyl azide [32]. In the middle of the 20th century, Huisgen systematically studied the family of 1,3-dipolar cycloaddition reactions [33], and henceforth the reaction is known as Huisgen reaction. However, the non-catalyzed Huisgen reaction of azide–

Table 1 Comparison of the reaction rate for different cycloaddition reactions

Reaction name	Reaction scheme	Reaction rate ($M^{-1}s^{-1}$)
Cu(I)-catalyzed azide - alkyne cycloaddition (CuAAC)	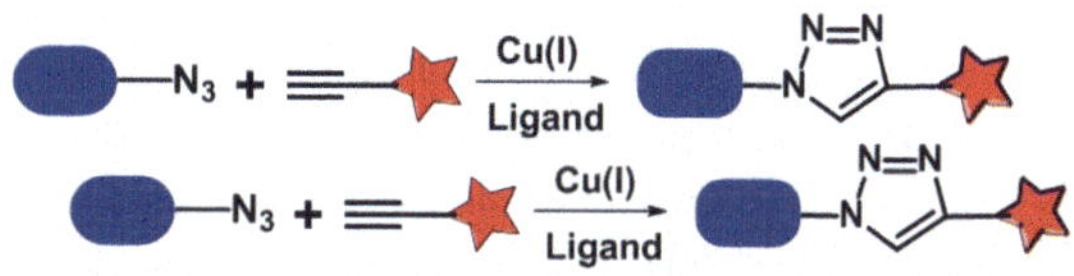	10–100 [15–18]
Strain-promoted azide–alkyne cycloaddition (SPAAC)	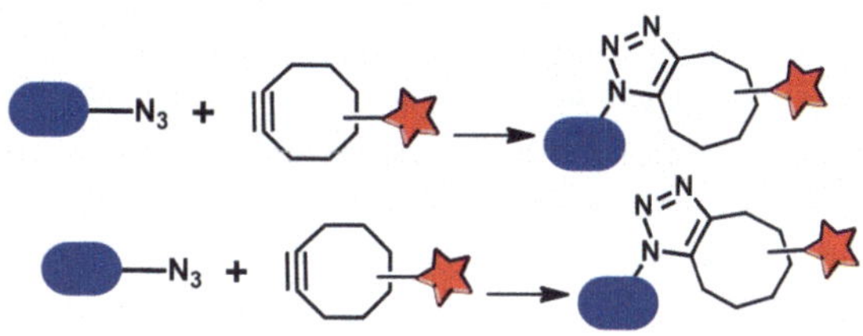	10^{-2}–1 [8, 19–22]
Inverse electron-demand Diels–Alder cycloaddition (IEDDA)	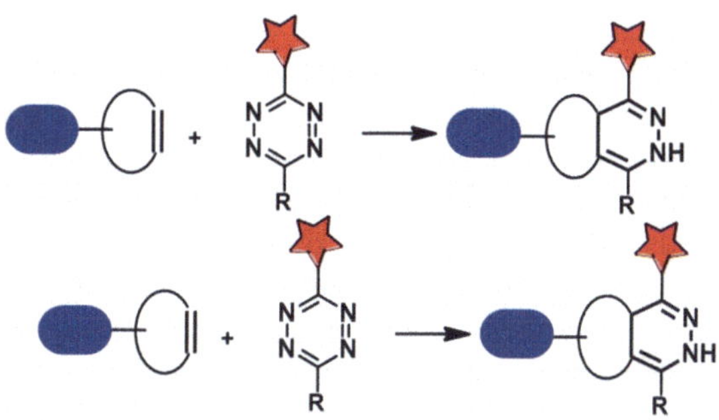	1–10^{4} [9, 10, 23, 24]
Photo-induced tetrazole-alkene cycloaddition (Photoclick)	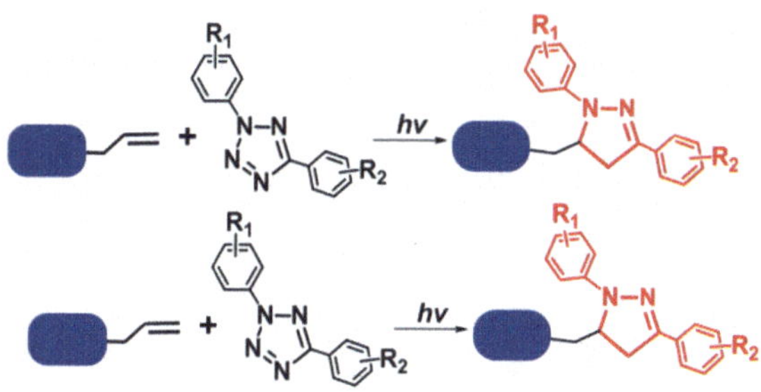	0.1–10^{3} [25–28]

alkyne cycloaddition produced a mixture of 1,4- and 1,5-disubstitution products. In addition, this reaction needs to be proceeded at high temperature and the reaction rate is very low, which hinders its wide applications.

In 2002, two research groups independently reported the efficient Cu(I) catalysis of the azide–alkyne cycloaddition (CuAAC) (Rostovtsev et al. [34] and Tornøe et al. [35]), which immediately drew wide attention from researchers all over the world. The CuAAC reaction represents the prototypical and the most commonly used 'click reaction', a term coined by Kolb et al. [36]. Since then, it has been widely adopted in synthetic chemistry, and material science, as well as in biomedical research [37–39]. In particular, the bioorthogonal nature of the azide and alkyne groups lends the CuAAC reaction as an invaluable tool for selective labeling and manipulation of intact biomolecules in living systems. The success of the CuAAC reaction can be attributed to the following three aspects: (i) the azide and alkyne functional groups are mutually reactive but remain inert to other chemically active groups found in living species; (ii) the reaction can proceed very fast, even in living systems, with a second-order rate constant of 10–100 $M^{-1}s^{-1}$ (Table 1); and (iii) the only product of this reaction is 1,4-triazole, which is regioselective, stable, and unreactive towards other active molecules.

Since Cu(I) species can accelerate the rate of the azide–alkyne cycloaddition reaction by more than seven orders of magnitude compared with the reaction without metal catalysts [40], its role in catalyzing the triazole formation has been the subject of intensive studies. Based on density functional theory (DFT) calculations, Himo and coworkers described the catalysis to be mediated by a single Cu(I) ion. They further concluded that the rate enhancement was due to a stepwise process that lowered the transition state energy approximately 11 kcal/mol compared with the uncatalyzed concerted cycloaddition [40]. Although these calculations brought insights into the catalytic cycle, the transition state remained controversial because the kinetic study indicated that the rate of the reaction was at least second order with respect to the concentration of Cu(I) ions. It is likely that more than 1 Cu atom is directly involved in formation of the transition state of this reaction. In a recent study, Worrell and coworkers provided direct evidence suggesting the presence of a dinuclear copper intermediate during the CuAAC reaction [41] (Fig. 1). Their

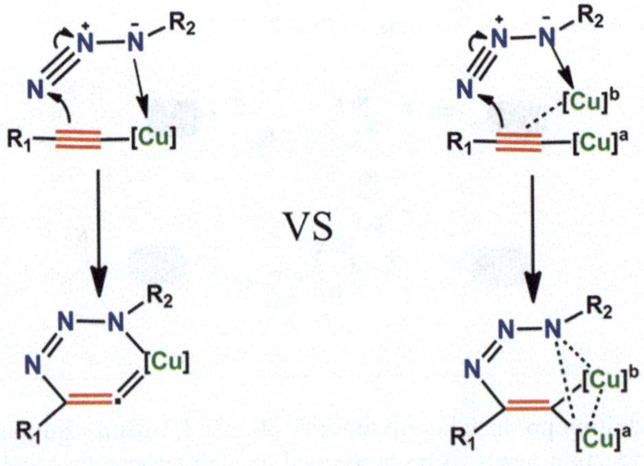

Fig. 1 Proposed intermediates of CuAAC with one or two copper atoms [41]. *CuAAC* copper(I)-catalyzed azide–alkyne cycloaddition

proposed bimetallic mechanism was supported by two experimental observations: (i) by using heat-flow reaction calorimetry to monitor the cycloaddition process in real-time, the authors found that the monomeric copper acetylide complex was not reactive unless an exogenous copper catalyst was added; (ii) in the crossover experiments with an isotopically enriched exogenous copper source, the direct injection time-of-flight mass spectrometry (TOF–MS) illustrated the stepwise nature of the carbon–nitrogen bond-forming events and the presence of two equivalence of copper atoms within the cycloaddition steps. Nevertheless, the real copper intermediate in catalyzing CuAAC remains to be further clarified.

A wide range of research articles have been published since the discovery of the CuAAC reaction, which underscore the robustness, broad applicability, and biocompatibility of this reaction. The choice of catalyst is largely dictated by the particular requirement of the experiment, especially when it comes to biological samples. Different copper(I) sources can be used in this reaction [42]; however, considering the solubility in aqueous buffer and toxicity to living cells, the most commonly used Cu(I) source in bioconjugation is the Cu(II) sulfate in the presence of reducing agents. As a mild reductant, ascorbate was introduced by Rostovtsev and coworkers to convert Cu(II) to Cu(I) [34], which, in conjunction with Cu(II) sulfate, has been wildly accepted as the choice of catalyst for CuAAC reaction performed on biomolecules.

2.2 Ligand-Assisted CuAAC Reaction

Although ligands are usually not required to carry out the CuAAC reaction, they can significantly enhance the reaction rate. Due to the oxidation of Cu(I), the reaction rate can be low if the catalyst is not presented at high concentration, which could be a problem for bioconjugation. The effect of ligands in CuAAC reaction can be due to its direct influence on the stability of the catalytic complex. For example, the ligand can be coordinated to Cu(I) ion during the reaction cycle, thus protecting it from oxidation. Some of the most commonly used ligands for CuAAC-mediated bioconjugation are listed in Table 2 and are discussed below.

The bipyridine-type compounds have been shown to be effective accelerating ligands with excellent solubility in water. By screening a number of bipyridine-type compounds, Lewis and coworkers discovered a highly active catalyst Cu(I)–sulfonated bathophenanthroline (BPS) [16]. Different ratios of BPS:Cu(I) catalysts were prepared to study the kinetics of the reaction. A second-order rate constant was observed that peaked at the 2:1 ratio between BPS and Cu(I), suggesting that two equivalents of ligand were bound to one equivalent of copper ions, and extra ligands would inhibit this reaction. To demonstrate the preparative and bioconjugation applications of this system, Gupta and coworkers used a BPS-Cu(I)-catalyzed CuAAC reaction on the cowpea mosaic virus (CPMV) particle [5]. Azide- or alkyne-modified sugars, peptides, and polymers, as well as fluorophores, have all been specifically conjugated to CPMV particles with high labeling yields. However, the BPS-Cu(I) complex is strongly electron rich and thus highly sensitive to oxygen; therefore, this system is usually performed in an inert environment that is not convenient for many biological applications.

Table 2 List of accelerating ligands for CuAAC reaction

Name/abbreviation	Structure	Characteristics	Applications
Bathophenanthroline disulfonate disodium salt (BPS)	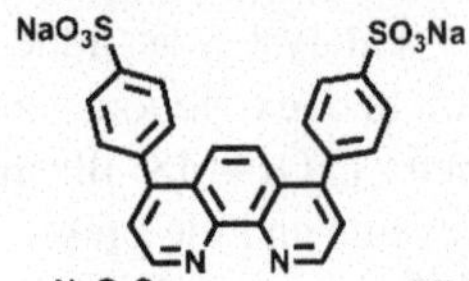 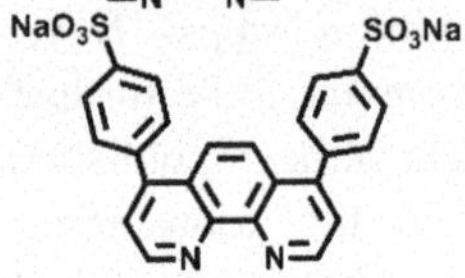	Water soluble; low membrane permeability	Proteins in vitro [45]; virus [5]
Tris-[(1-benzyl-1H-1,2,3-triazol-4-yl)methyl]amine (TBTA)	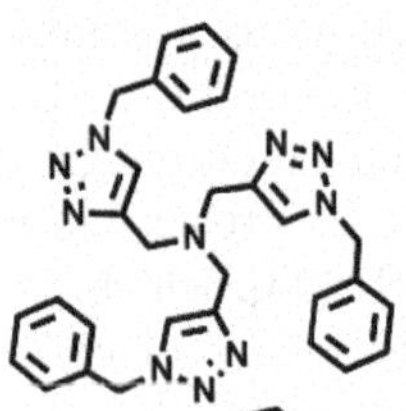 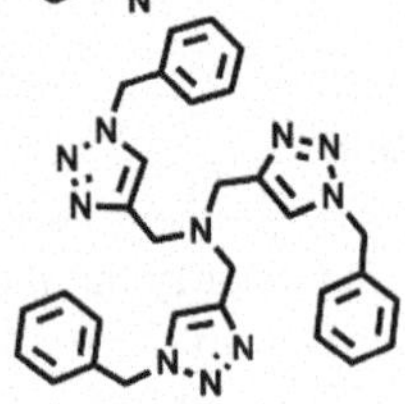	Poor water solubility; cell toxic	Proteins in vitro; *E. coli* membrane protein [46]
Tris[(1-hydroxypropyl-1H-1,2,3-triazol-4-yl)methyl]amine (THPTA)	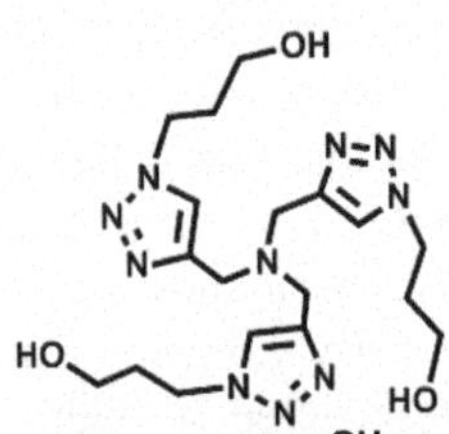 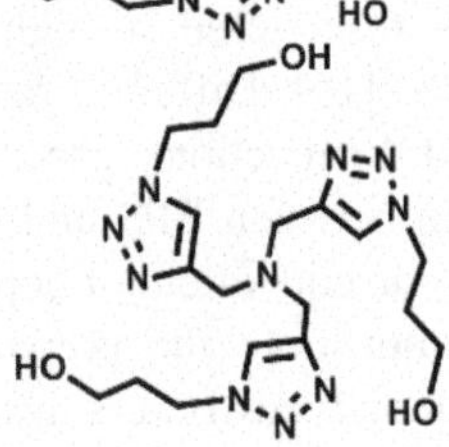	Water soluble	Live cell surface glycans [47]

Table 2 continued

Name/abbreviation	Structure	Characteristics	Applications
2-[4-{(bis[(1-tert-butyl-1H-1,2,3-triazol-4-yl)methyl]amino)methyl}-1H-1,2,3-triazol-1-yl]-acetic acid (BTTAA)	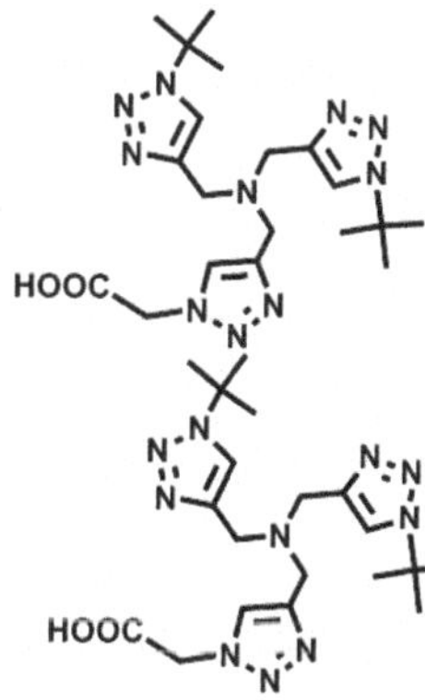	Water soluble	Live cell surface glycans [48]; *E. coli* periplasm proteins [49]
3-[4-({bis[(1-tert-butyl-1H-1,2,3-triazol-4-yl)methyl]amin}methyl)-1H-1,2,3-triazol-1-yl]propanol (BTTP)	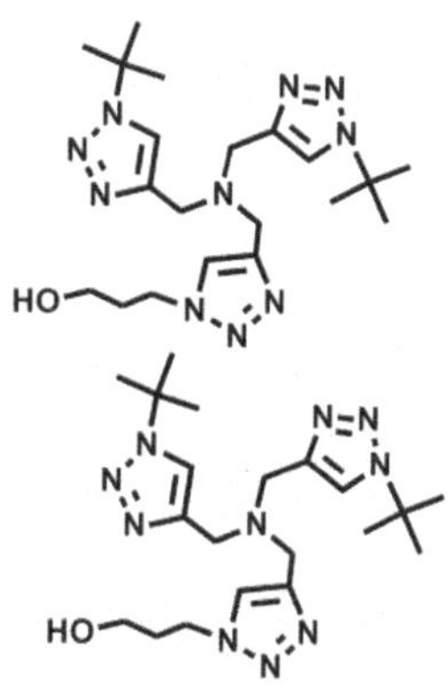	Water soluble; membrane permeable	Live cell surface glycans [50]; *E. coli* periplasm and cytoplasm proteins [51]
2-[4-{(bis[(1-tert-butyl-1H-1,2,3-triazol-4-yl)methyl]amino)-methyl}-1H-1,2,3-triazol-1-yl]ethyl hydrogen sulfate (BTTES)	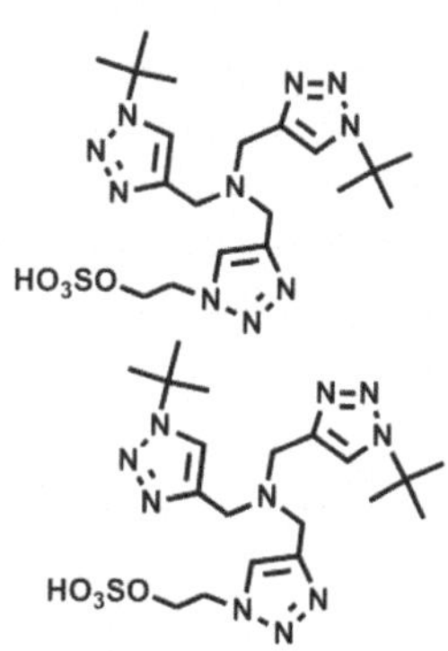	Water soluble; negative charge; low membrane permeability	Live cell surface glycans [52]

Table 2 continued

Name/abbreviation	Structure	Characteristics	Applications
3-[4-({bis[(1-tert-butyl-1H-1,2,3-triazol-4-yl)methyl]amino}methyl)-1H-1,2,3-triazol-1-yl]propyl hydrogen sulfate (BTTPS)	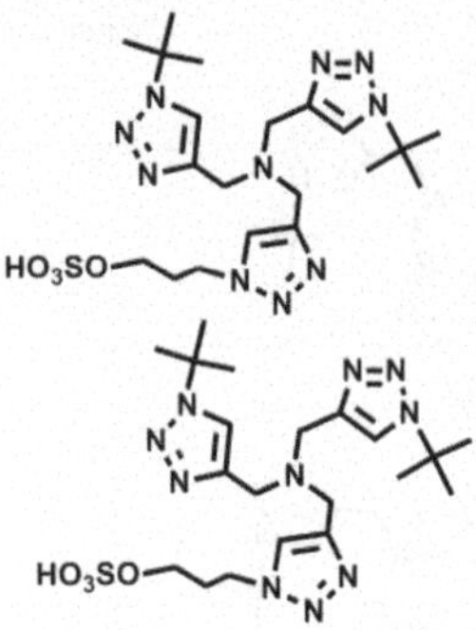	Water soluble; negative charge; low membrane permeability	Live cell surface glycans; *E. coli* membrane protein [50]
L-histidine (His)	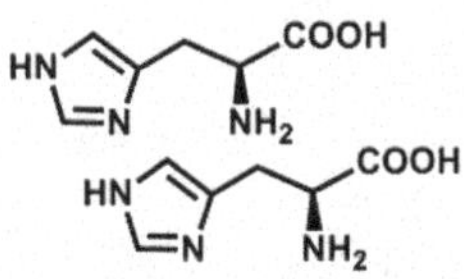	Water soluble; membrane permeable	Live cell surface glycans [53]

To date, the most widely used accelerating ligands in the bioconjugation field were based on tris(triazolylmethyl)amine, which themselves were prepared using the CuAAC reaction. During the synthesis of polytriazole compounds, Chan and coworkers found that the reaction rate was higher than usual and it appeared to be autocatalytic. They thus hypothesized that the polytriazole products themselves could potentially serve as accelerating ligands for CuAAC reaction [43]. Their subsequent study indicated that these oligotriazole ligands were effective in protecting Cu(I) ions from aerobic aqueous conditions and increasing the reaction rate. Rodionov and coworkers further investigated the molecular details of these triazole ligands in accelerating the CuAAC reaction, and a possible mechanism was proposed [44]: the central nitrogen donor provides electron density to the Cu(I) atom; the three-armed motif, carrying relatively weak coordinating heterocyclic ligands, binds to the Cu(I) atom with sufficient strength while providing access to necessary coordination sites.

As the first generation of triazole ligands that have been used for bioorthogonal labeling, TBTA significantly accelerates the reaction and stabilizes the Cu(I) oxidation state in aqueous buffer. Its application was first demonstrated by the efficient attachment of different alkyne-containing fluorescent molecules to the azide-labeled CPMV particles [5]. Later, Link and Tirrell performed TBTA-assisted CuAAC reaction on the surface of living *Escherichia coli* cells. They incorporated the alanine analog azidohomoalanine (AHA) into the *E. coli* outer membrane protein C (OmpC), followed by the CuAAC labeling with an alkyne-biotin probe [46]. These selectively labeled *E. coli* cells could be further stained with a fluorescent avidin molecule and specifically detected by flow cytometry; however, these *E. coli* cells were found to be unable to divide after being transferred to rich medium [54], indicating the severe toxicity of the TBTA-Cu(I) complex to living cells. Moreover, the poor solubility of TBTA in aqueous buffer further hindered its applications in biological systems.

To circumvent these problems, Hong and coworkers developed a more polar and less toxic ligand, THPTA, which showed excellent efficiency in assisting the CuAAC-mediated labeling of CPMV particles [55]. They further applied the THPTA-assisted CuAAC reaction to label the azide-containing glycans with an alkyne-modified fluorophore on the surface of HeLa, CHO and Jurkat cells, which all yielded significant fluorescence signal. This work represents the first example of ligand-assisted CuAAC labeling reaction on a mammalian cell surface [47]. In the same study, the authors found that THPTA prevented the toxic effects of Cu(I) ions in a dose-dependent manner. A 5:1 ratio of THPTA:Cu(I) was found to preserve the cell viability at copper concentrations ranging from 10 to 50 μM.

Two novel TBTA analogs, BTTES and BTTAA, were recently developed by Besanceney-Webler et al. [48, 52] and Soriano del Amo et al. [48, 52]. Ligand-assisted CuAAC reactions were compared side-by-side in the presence of TBTA, THPTA, BTTAA, and BTTES. Labeling efficiency was measured by fluorescent imaging of fucosylated glycans on the cell surface, which showed that the newly developed TBTA analogs (BTTAA and BTTES) accelerated the reaction more efficiently than other ligands. Furthermore, they demonstrated that 50 μM of BTTAA-Cu(I) complex showed negligible influence on cell viability, while cells treated with the same amount of TBTA-Cu(I) complex exhibited a slower rate of proliferation. These results suggested that the new ligands could effectively suppress Cu(I)-produced toxicity, making it a better choice for conducting CuAAC-mediated cell surface labeling. Most recently, Wang and coworkers developed two additional TBTA analogs, BTTP and BTTPS [50], both of which could effectively increase the reaction rate while further eliminating the toxicity of Cu(I). For example, the negatively charged sulfate group on BTTPS may decrease the membrane permeability of the entire Cu(I)-ligand complex, which could help reduce the toxicity and make it a better choice for labeling biomolecules on the cell surface.

Another interesting study by Kennedy and coworkers demonstrated that the natural amino acid L-histidine (His) could serve as the Cu(I) ligand that effectively catalyzed the CuAAC labeling of alkyne-tagged glycans on the mammalian cell surface [53]. The authors analyzed the cytotoxicity of this copper complex by measuring mitochondrial activity using the MTT assay [MTT: 3-(4,5-dimethylthiazol-2-yl)- 2,5-diphenyltetrazolium bromide], which showed that this copper complex had no apparent toxicity to all cell lines tested. Considering the high cell permeability of histidine, the authors suggested utilization of this simple ligand for assisting CuAAC reaction inside living cells.

2.3 Chelation-Assisted CuAAC Reaction

In 2009, Brotherton and coworkers reported that certain azide substrates containing auxiliary nitrogen donor ligands could effectively accelerate the CuAAC reaction, which they termed as 'chelation-assisted CuAAC reaction' [56]. Further mechanistic study revealed that chelation-assisted binding between the alkylated azido nitrogen (N-α) and the catalytic copper center greatly enhanced the electrophilicity of the azido group, which then lowered the kinetic barrier to form the key metallacycle intermediate upon nucleophilic attack by the copper acetylide [57] (Fig. 2a). After initial demonstration in organic solvents, Uttamapinant and coworkers later applied this chelation-assisted CuAAC reaction for biomolecule labeling [18]. The authors fused the lipoic acid ligase (LplA) acceptor peptide (LAP) to a target protein on cell surface, followed by the enzymatic conjugation with picolyl azide substrate. By treating live cells with an alkyne-modified Alex647 fluorophore in the presence of Cu(I), significant fluorescence signal was observed in the positively labeled cells. Moreover, the authors found that the use of picolyl azide, in conjunction with THPTA or BTTAA ligand, could increase the specific protein labeling signal by as much as 25-fold in comparison with the conventional non-chelating azide. To further demonstrate the biocompatibility of this new reaction system, they applied this strategy to label the postsynaptic transmembrane protein neuroligin-1 on the surface of neuron cells (hippocampal rat neurons) (Fig. 2b). Together, the use of chelation-assisted CuAAC reaction effectively decreased the Cu(I) concentration from 100 to 10 μM, and showed negligible toxicity to neuron cells without sacrificing the labeling efficiency.

Recently, Jiang and coworkers reported that a modified picolyl azide, bearing an electron-donating group rather than the previous electron-withdrawing group, could

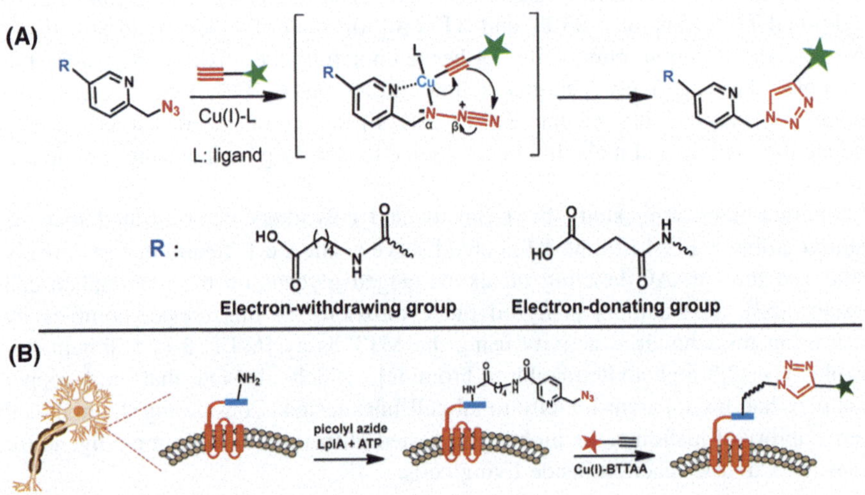

Fig. 2 Chelation-assisted CuAAC reaction. **a** Schematic illustration of the chelation model of picolyl azide–alkyne cycloaddition. **R** group can be substituted by either the electron-withdrawing or electron-donating group; **b** Selective labeling of neuron cell-surface proteins with an engineered picolyl azide ligase and chelation-assisted CuAAC [18, 58]. *CuAAC* copper(I)-catalyzed azide–alkyne cycloaddition

further increase the reaction rate [58] (Fig. 2a). The authors found that this newly designed azide substrate could boost the efficiency of ligand-assisted CuAAC reaction by 20- to 38-fold in the living system without apparent toxicity. With a combination of this unique azide and BTTPS ligand, they detected newly synthesized cell-surface glycans by flow cytometry analysis. They further used this supersensitive chemistry to monitor the dynamic glycan biosynthesis in mammalian cells as well as during the early embryogenesis of zebrafish. In addition, the same work also discovered that it took approximately 30–45 min for a monosaccharide building block to be metabolized and incorporated into cell-surface glycoconjugates.

3 Biological Applications of CuAAC Reaction

Selective labeling of biomolecules in living systems is essential to understand the complex nuances of spatial–temporal aspects of biomolecule function and localization [13, 59–61]. To date, CuAAC has proven to be effective and the most frequently used reaction for this challenging task, mainly due to three reasons: (i) it is biocompatible and the reaction rate is very high; (ii) the azide and alkyne groups are more synthetically accessible; and (iii) the size of the azide and alkyne groups is small enough to minimize the perturbation when incorporated into the biomolecules. It has therefore been widely employed for selective labeling and visualization of a wide range of biomolecules, including proteins, sugars, nucleic acids and lipids [13, 62, 63], which will be discussed in detail below. There have been many in-depth reviews on these topics [29, 59, 64, 65] and here we only introduce a few selected examples to highlight the broad utility of CuAAC reaction.

3.1 Applications on Proteins

Proteins are the most abundant biomolecules in living systems and participate in almost all biological processes. Protein labeling has become a powerful tool not only for protein engineering (e.g. PEGylation) but also for the study of their localization, trafficking and function, particularly within living cells [66–68]. CuAAC has emerged as the most frequently used bioorthogonal reaction for protein labeling. As a prerequisite for CuAAC-mediated protein labeling, a variety of strategies have been developed for selective incorporation of the alkyne or azide handles into proteins, including residue-specific incorporation, enzymatic modification and unnatural amino acid (UAA) mutagenesis [69–76].

Link and Tirrell reported the first CuAAC reaction for membrane protein (OmpC) labeling on living *E. coli* cells using a metabolic incorporated UAA (AHA; Fig. 3) [46]. The alkyne-containing UAA homopropargylglycine (HPG) was later developed, which, in conjunction with TBTA-assisted CuAAC reaction, allowed fluorescent labeling and visualization of newly synthesized proteins in proteome level in both *E. coli* and mammalian cells [77, 78] (Fig. 3). Recently, the same group expanded this strategy to rat hippocampal neuron tissues. Regulation of protein synthesis in dendrites is important for neuronal synapses responding

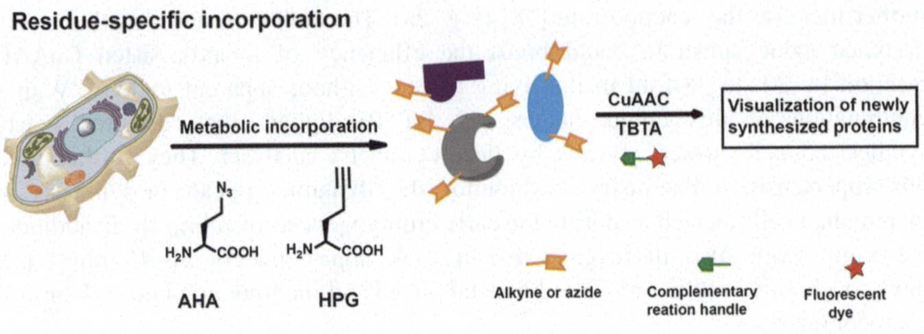

Fig. 3 Coupling the residue-specific UAA incorporation with CuAAC reaction for labeling and visualization of newly synthesized proteins in situ [77, 78]. *UAA* unnatural amino acid, *CuAAC* copper(I)-catalyzed azide–alkyne cycloaddition, *AHA* azidohomoalanine, *HPG* homopropargylglycine, *TBTA* tris-[(1-benzyl-1H-1,2,3-triazol-4-yl)methyl]amine

dynamically and specifically to stimulations [79, 80]. Combining CuAAC reaction and pulse-chase incorporation of AHA or HPG allowed the visualization of protein synthesis in different time periods, as well as different cell components in hippocampal neurons [81, 82]. Furthermore, they used a microperfusion technique to selectively incorporate the AHA (or AHA together with a protein synthesis inhibitor) in somata and dendrites, respectively, and observed the spatial origin of the newly synthesized proteins in both cell bodies and dendrites.

The coupling of residue-specific UAA incorporation and CuAAC reaction provides a way for visualization and/or identification of the general proteomic responses. In a complementary strategy, the combination of enzymatic modification with CuAAC reaction provides valuable tools for site-specific protein labeling in living systems. For example, Uttamapinant and coworkers integrated the LplA-mediated enzymatic modification with ligand-assisted or chelation-assisted CuAAC reaction to realize site-specific labeling of postsynaptic transmembrane protein neuroligin-1 on living mammalian cell surface [18] (Fig. 2b). For such enzymatic modifications, the incorporation sites of the peptide tags are still largely limited to the N and C terminal or the internal loops [71]. In contrast, the incorporation sites for azide and alkyne handles are less restricted during UAA mutagenesis, which makes it a powerful tool for site-specific protein labeling.

Recently, our group employed the BTTAA-assisted CuAAC reaction to label an acid-responsive chaperone protein (HdeA), affording a protein-based pH sensor for measuring the extremely low pH values that enteric bacteria will encounter when passing through human stomach [49] (Fig. 4). An azide-bearing UAA (ACPK) was genetically and site-specifically incorporated into HdeA which can adopt pH-dependent graduate unfolding during acidification. The alkyne-functionalized environment-sensitive fluorophore (alk-4-DMN) was then site-specifically conjugated to HdeA via the BTTAA-assisted CuAAC reaction through this ACPK handle. In vitro survey of the attachment sites by fluorescence spectroscopy led to the discovery of one position (Val 58) on HdeA that showed a significant fluorescence increase upon acidification, making it possible to act as a pH indicator for the acidic environment. Indeed, *E. coli* cells harboring this HdeA-4-DMN

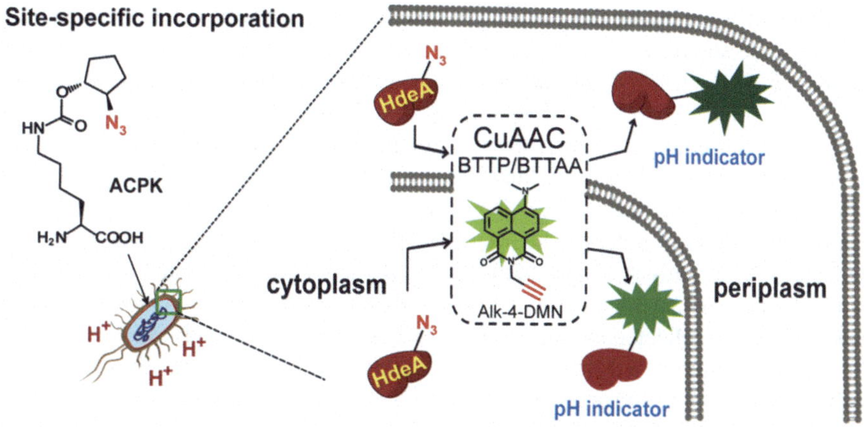

Fig. 4 Coupling site-specific UAA incorporation with CuAAC reaction allowed the development of a protein-based pH indicator for bacterial periplasmic and cytoplasmic spaces [49, 51]. *UAA* unnatural amino acid, *CuAAC* copper(I)-catalyzed azide–alkyne cycloaddition, *ACPK* N^{ε}-(((1R,2R)-2-azidocyclopentyloxy)carbonyl) -L-lysine

variant showed strong fluorescent enhancement when the environmental pH dropped from 7 to 2.3.

By collaborating with the Zhao et al. group and Wu et al. group, we further expanded the ligand-assisted CuAAC labeling into bacterial cytosolic space [51]. By systematic surveying, the labeling efficiency of a panel of Cu(I) ligands, including BPS, TBTA, BTTAA, BTTP, BTTPS and His, we found that BTTAA and BTTP could effectively enhance CuAAC-mediated labeling of proteins in bacteria cytosol (Fig. 4). Toxicity study showed that the ligand BTTP significantly increased the redox potential of the coordinated Cu(I) ions, which effectively sequestered Cu(I)-associated toxicity while accelerating the reaction rate. We specifically directed our previously developed pH indicator to both *E. coli* periplasm and cytoplasm and compared the acid tolerance of these different intracellular spaces. The results showed that when the extracellular pH dropped to 3, the cytoplasm- and periplasm-specific pH measurement showed an approximately 1.1 pH unit trans-membrane gradient; this value is important when studying bacterial acid-resistance mechanism.

3.2 Applications on Glycans

Glycosylation is an important post-translational modification on proteins and its physiological and/or pathological roles during immune response, inflammation, and cancer metastasis, as well as cell surface recognition, remain to be clarified [83, 84]. Spurred by the flexible substrate tolerance during glycan biosynthesis, metabolic incorporation of sugars with slight structural modifications onto glycoproteins has become a powerful strategy for glycan labeling, making it an important tool to study these complicated biological systems [85, 86].

Employing the metabolic incorporation of unnatural sugars with CuAAC reaction, Jiang and coworkers recently reported the dynamic single molecule tracking and super-resolution imaging of glycans on cancer cell surface, which may aid the understanding of the roles of surface glycan during cancer growth and metastasis. [87]. In this work, alkyne- or azide-tagged glycans were metabolically incorporated into the peripheral position of both N-linked sialic acids and O-linked acetyl galactosamie (GalNAc), permitting the BTTPS-assisted CuAAC labeling with azide- or alkyne-bearing fluorophores (Fig. 5). To reduce the high spatial density, part of the fluorescent molecules were bleached through a series of controlled excitation pulses, which enabled the dynamic tracking of single glycan molecules. This study revealed that glycans diffused more slowly in those cells that had a higher metastatic potential compared with cells with a lower metastatic potential, which reflected the increased crowding of the tethered glycan polymers on the surface of highly metastatic cells. Furthermore, using this labeling scheme, the authors labeled cell surface glycans with red cyanine fluorophores, making it suitable for super-resolution imaging such as stochastic optical reconstruction microscopy (STORM).

In addition to imaging applications, CuAAC was also applied for profiling glycoproteomics and studying their function (Fig. 5). Sialic acids belong to the family of nine-carbon monosaccharides, and hypersialylation has been reported to increase the mobility of cancer cells [88, 89]. By using alkyne-tagged sugars in combination with an azide-tagged biotin probe and the CuAAC reaction, Liu and coworkers identified 87 and 144 sialylated proteins in two different lung cancer cell lines with distinct invasiveness (Cl1-0 and Cl1-5), respectively [90]. Among these sialylated proteins, epidermal growth factor receptor (EGFR) showed higher sialylation and fucosylation levels in Cl1-5 (more invasive) than in Cl1-0. Through further study, the authors found that increasing sialylation and fucosylation would suppress EGFR dimerization and phosphorylation upon EGF treatment, which attenuated the EGFR-mediated invasion of lung cancer cells.

Recently, Du and coworkers studied the dynamic sialylation in regulating epithelial-mesenchymal transition (EMT) [91]. EMT is a fundamental process during embryonic development and organ formation, and aberrant regulation of EMT often leads to tumor progression. These facts suggest a correlative link between EMT and hypersialylation. To address the dynamic changes of sialylation during EMT and better understand the contribution of glycosylation in regulating

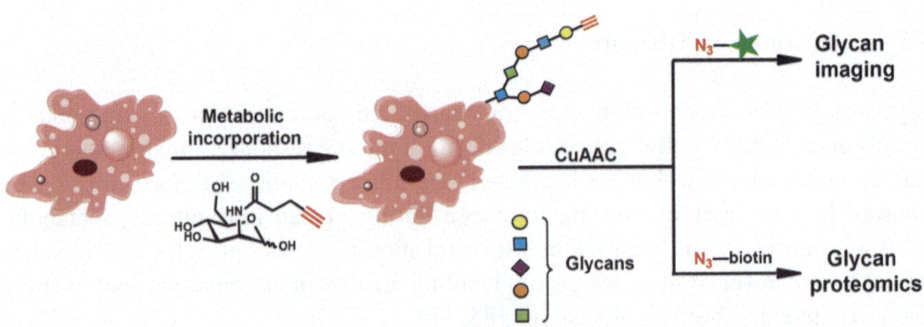

Fig. 5 Metabolic incorporation of alkyne-tagged sugars in conjunction with CuAAC reaction for glycan imaging and glycoproteomic profiling [87, 90]. *CuAAC* copper(I)-catalyzed azide–alkyne cycloaddition

cell pathology and physiology, an azide-containing sialylated glycan was metabolically incorporated on the cell surface, followed by the BTTAA-assisted CuAAC labeling with an alkyne-modified fluorophore. Dynamic imaging revealed that sialylation was downregulated during EMT but was then reverted and upregulated in the mesenchymal state after EMT. These observations indicated a correlation between glycosylation and the multistep progression through EMT. The sialylated proteins were further pulled down via CuAAC labeling with biotin alkyne and sent for mass spectrometry (MS) analysis. This quantitative proteomic analysis identified a list of sialylated proteins with relative abundance being either up- or downregulated. Together, these findings verified the important roles of sialylation in regulating EMT, and indicated its possible function in related pathological events such as cancer metastasis.

More recently, Woo and coworkers developed a mass-independent platform for intact N- and O-glycopeptide discovery and analysis in glycoproteomics [92]. Due to the non-uniformity nature of protein glycosylation, intact glycosylated peptides remain difficult for MS analysis as well as the database search. In their strategy, azide-tagged sugars were first metabolically incorporated into proteins; a multifunctional probe containing an alkyne moiety, a biotin moiety, an isotope recoding moiety, and a cleavage moiety was then labeled to the glycosylated proteins through BTTP-assisted CuAAC reaction. After biotin enrichment, trypsin digestion and cleavage, the glycopeptides containing the isotope recode moiety were released from the beads and analyzed by liquid chromatography–mass spectrometry (LC–MS). Through an isotope pattern-searching algorithm developed in their previous work, the glycopeptides could be directly picked up from the backgrounds for tandem MS (i.e. MS2 and MS3) analysis, which led to high confidence glycopeptide identification (Fig. 6). Using this platform, they assigned 32 N-linked glycopeptides and over 500 O-linked glycopeptides from a total of 250 proteins.

3.3 Applications on Lipids

Lipids are the most abundant metabolites in cells, which comprise a variety of species, including phospholipids, phosphatidylinositol phosphates (PIPs) and cholesterol [93]. These molecules are not only the building blocks for cell membrane but are also involved in various cellular processes, including signaling transduction, protein trafficking, modifications, interactions and regulation [94]. Combining CuAAC reaction with azide- or alkyne-functionalized lipid analogs provides a facile way for studying these processes [95].

Jao and coworkers developed a choline (Cho) analog propargylcholine (containing an alkyne group, propargyl-Cho) for in vivo visualization and tracking of Cho-containing phospholipids [96]. After cell fixation, propargyl-Cho was labeled with two different reagents (Alexa568-azide and biotin-azide) through CuAAC reaction. The biotin tag was further detected by Alexa488-streptavidin, which is membrane impermeable. This strategy enabled the simultaneously detection of Cho-containing phospholipids in different cell components. The authors further expanded the labeling of propargyl-Cho to analyze the phospholipid synthesis in different organs in mice.

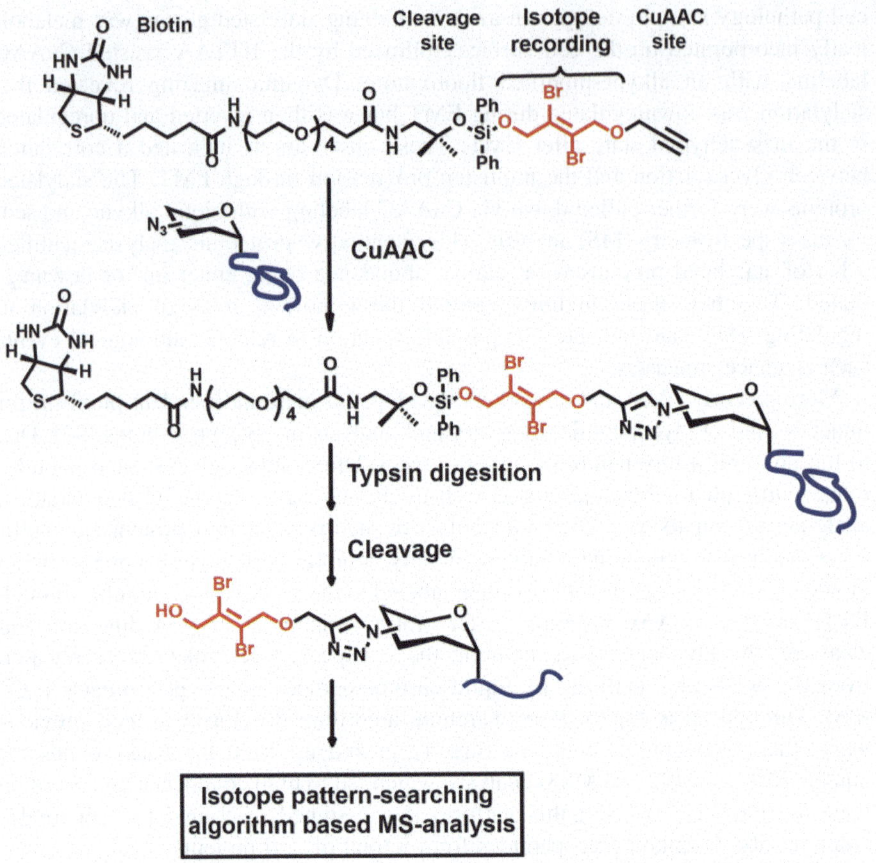

Fig. 6 Schematic representation for mass-independent identification of glycoproteomics using an alkyne-tagged multifunctional probe and the CuAAC reaction [92]. *CuAAC* copper(I)-catalyzed azide–alkyne cycloaddition

Hang and coworkers developed several azide- or alkyne-tagged lipid probes for studying protein lipidation [95]. For example, the authors designed an alk-16 palmitate reporter for profiling S-palmitoylated proteins [97] (Fig. 7a). Among the identified 150 proteins with diverse cellular functions, the authors took interferon-induced transmembrane protein 3 (IFITM3) for further study and found that S-palmitoylation was modified at three membrane-proximal residues: C71, C72 and C105. This S-palmitoylation on IFITM3 was found to control its clustering in membrane compartments as well as the antiviral activity against influenza virus. In addition to covalent lipid modifications, Hang group as well as other groups, also developed a serious of bifunctional probes for investigating non-covalent lipid-protein interactions as well as lipid-mediated protein–protein interactions [63, 99, 100]. These bifunctional probes typically contain a photo-affinity moiety and an alkyne or azide group. Via ultraviolet irradiation, the interacting proteins on lipids can be covalently captured by the photo-affinity moiety, while the alkyne or azide

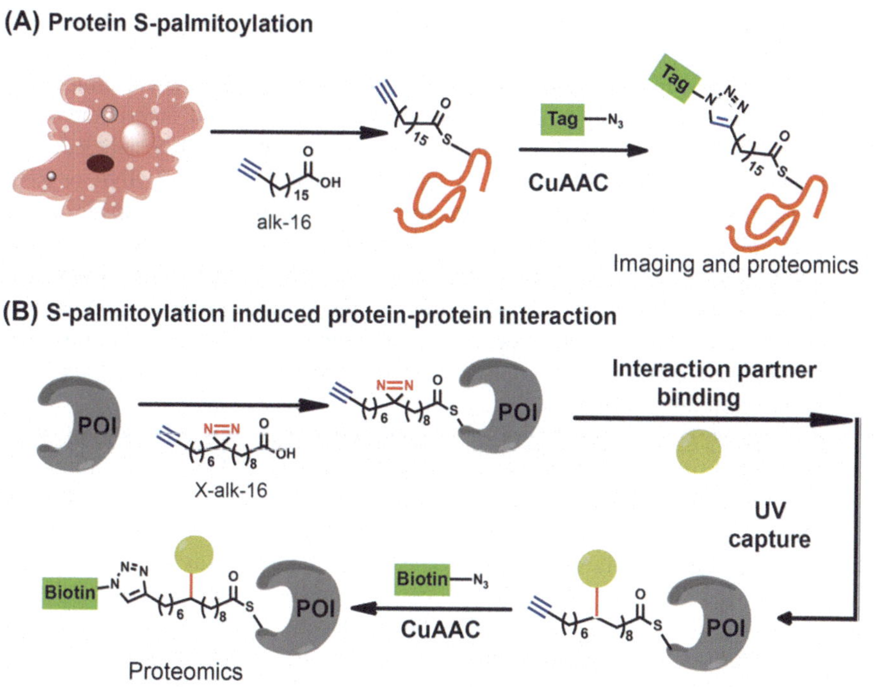

Fig. 7 CuAAC reaction for studying protein lipidation and lipidation-induced protein–protein interactions. Schematic representation for studying (**a**) protein S-palmitoylation using an alkyne-tagged lipid probe alk-16 and the CuAAC reaction, and (**b**) S-palmitoylation-induced protein–protein interactions using a multifunctional lipid probe X-alk-16 [97, 98]. *CuAAC* copper(I)-catalyzed azide–alkyne cycloaddition, *UV* ultraviolet

handle can be subsequently labeled with a fluorescent probe or enriched by a biotin tag through CuAAC reaction. To this end, Hulce et al. developed a panel of bifunctional cholesterol probes containing both alkyne and diazirine moieties [101]. In combination with quantitative MS, the authors identified over 250 cholesterol-binding proteins, including receptors, channel proteins and enzymes. Recently, Peng and Hang reported a bifunctional fatty acid probe (X-alk-16) that contained an alkyne and a diazirine group. This probe can be metabolically incorporated into a variety of S-palmitoylated proteins such as IFITM3 and H-Ras, and be subsequently photocrosslinked with their interacting proteins [98] (Fig. 7b). The authors took IFITM3 for further study and identified 12 interacting proteins, including vesicle-membrane-protein-associated protein A (VAPA), an important partner for the full antiviral activity of IFITM3.

3.4 Applications on Nucleic Acids

Using bioorthogonal chemistry to label nucleic acids, including DNA and RNA, has been successfully performed and exploited in many studies [30]. Alkyne-modified nucleotides were first introduced into oligonucleotides through in vitro organic synthesis, which could then react with azide probes via CuAAC labeling. Salic and coworkers reported the CuAAC labeling of DNA and RNA, both in cultured cells and in mice [102, 103]. In their studies, a propargyl group was introduced onto deoxyuridine (5-ethynyl-2'-deoxyuridine, EdU) or uridine (5-ethynyluridine, EU), which was then uptaken by cellular biosynthesis machinery and incorporated into DNA or RNA molecules. These nucleotides were subsequently labeled with fluorescent tags via CuAAC reaction to achieve optical detection of DNA or RNA in vivo (Fig. 8a). Employing this strategy, the authors examined the turnover of bulk RNAs in cultured cells. They found that the fluorescent signal of RNA in nuclear decreased dramatically after initial staining, while the signal in the cytoplasm remained constant. Furthermore, they assayed the transcription rates in various mouse tissues in whole animals, both on sections and by whole-mount staining. The results indicated that the total transcription rates varied greatly among different tissues and different cell types.

Recently, Neef and coworkers developed a strategy for detection of virus infection by using a gemcitabine metabolite analog 2′-deoxy-2′,2′-difluoro-5-ethynyluridine (dF-EdU) and CuAAC reaction [104]. The incorporation of dF-EdU into DNA depends on its phosphorylation by thymidine kinase (TK). They found that dF-EdU was selectively phosphorylated by a herpes virus TK, but not the human TK (hTK). Only the infected cells expressing TK can metabolically incorporate dF-EdU into DNA and thus be labeled by azide-tagged fluorophores (Fig. 8b). This strategy offers a valuable tool for in situ detection of virus-infected host cells.

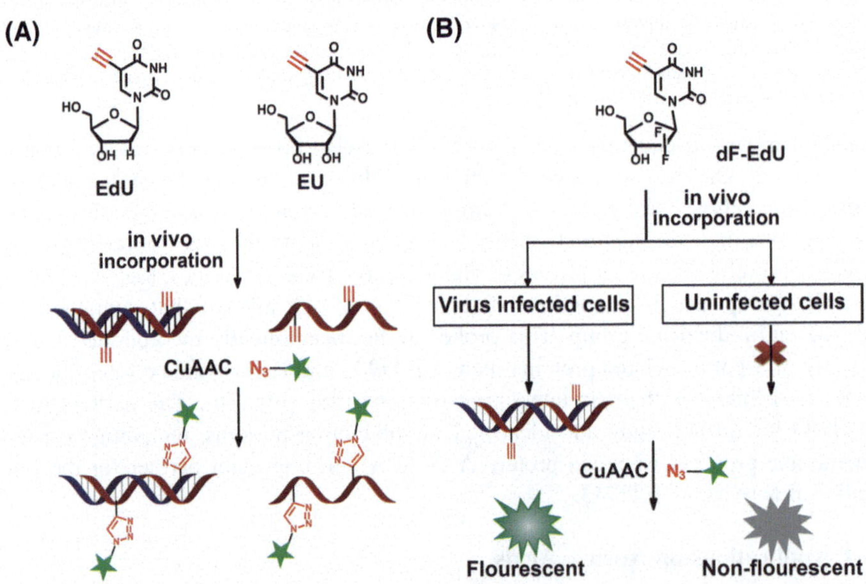

Fig. 8 CuAAC reaction for in vivo labeling of DNA and RNA. **a** Schematic representation for in vivo fluorescent labeling of DNA and RNA molecules using metabolically incorporated EdU (or EU) and CuAAC reaction. **b** Detection of virus infection using the cell-selective incorporation of dF-EdU and CuAAC reaction [102–104]. *CuAAC* copper(I)-catalyzed azide–alkyne cycloaddition, *EdU* 5-ethynyl-2-deoxyuridine, *EU* 5-ethynyluridine, *dF-EdU* 2′-deoxy-2′,2′-difluoro-5-ethynyluridine

3.5 Activity-Based Protein Profiling

In the post-genome era, assigning functions to thousands of newly predicted gene products remains a central task. Proteomics aim to accelerate this process by large-scale analysis of the entire proteome, particularly in a native biological system. However, due to the complexity of post-translational modifications, most of the proteomic techniques could not provide insights into the specific fraction of active proteins within a given system. To address this central problem, activity-based protein profiling (ABPP) was introduced to complement the existing genomic and proteomic strategies [105–107]. In a typical ABPP experiment, the activity-based probe (ABP) is comprised of a reporter group (e.g. fluorohpore or biotin) and a reactive electrophile for covalent modification of the active-site residue of proteins. The ABPs only label active enzymes within a protein mixture, and the labeled proteins can be visualized by in-gel fluorescence or identified using MS. These initially designed ABPs were bulky and hindered the cellular uptake and/or binding with target proteins when administered to live cells or organisms. This issue was recently circumvented by the functionalization of ABPs with sterically inconspicuous bioorthogonal handles, which could be modified with reporter groups via bioorthogonal chemistry ex vivo [108]. CuAAC has been one of the most frequently used bioorthogonal reactions for ABPP, which has helped to identify many previously unassigned active residues such as cysteine from diverse proteins (Fig. 9).

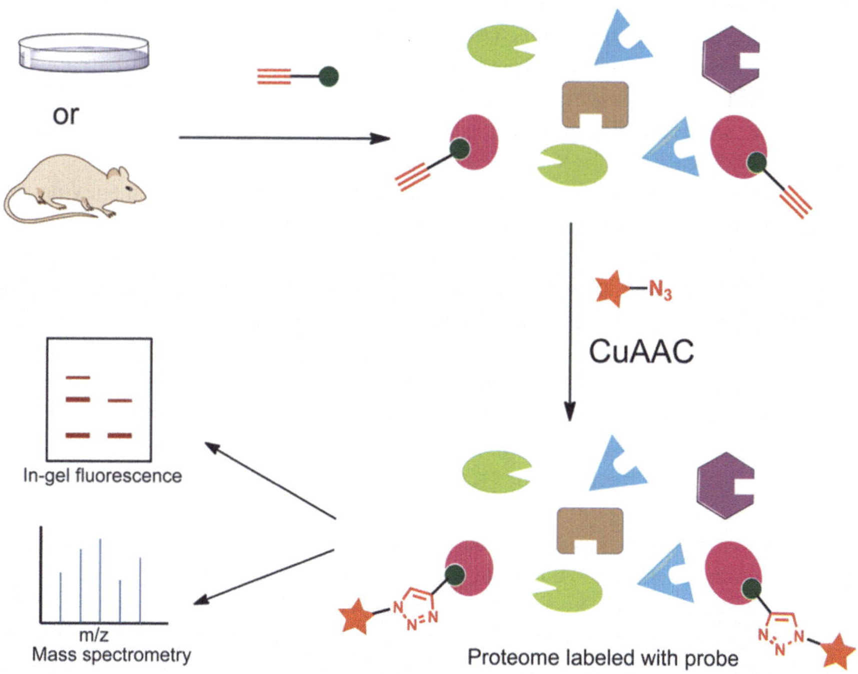

Fig. 9 Schematic representation for the ABPP strategy using CuAAC reaction. *CuAAC* copper(I)-catalyzed azide–alkyne cycloaddition, *ABPP* activity-based protein profiling

In particular, a variety of ABPs have been designed to probe enzyme classes, including hydrolases, kinases, phosphatases, glycosidases, histone deacetylases (HDACs) and cytochrome P450s [109, 110]. With the development of effective analytical techniques such as high-content mass spectrometric-based proteomic methods, ABPP has been further advanced and a number of interesting insights and applications have arisen from these works. Initially, the ABP-modified proteins were labeled with biotin through CuAAC reactions, and these enriched biotinylated proteins were then subjected to trypsin digestion and LC–MS analysis. However, this approach did not allow the identification of the modification sites because the ABP-attached peptide fragments were still conjugated to the streptavidin beads during enrichment. Speers and Cravatt later designed tandem-orthogonal proteolysis (TOP-ABPP) in which both the affinity tag (e.g. biotin) and the Tobacco Etch Virus (TEV) protease cleavage site were introduced to the ABP-labelled proteins via CuAAC labeling [111]. Furthermore, the same group introduced an isotopically labelled valine moiety into the TEV recognition site to facilitate the quantitative MS measurement, which was termed isoTOP-ABPP [112]. In this study, the authors first labeled active cysteine residues from the entire proteome with an alkyne-iodoacetamide (alk-IA) probe, and the light or heavy variants of the azide-TEV-biotin tags were then conjugated to proteins via CuAAC reaction and the samples were subjected to tandem on-beads proteolytic digestion with trypsin and TEV protease. The isotopic tag-labelled peptides were released and analyzed by LC–MS/MS, which could identify IA-modified cysteines and quantify the extent of labelling. The isoTOP-ABPP ratio, which was generated as the signal intensity between light and heavy tag-conjugated proteins, could be used to assess the intrinsic reactivity of cysteine residues within the native proteome. Notably, they found that the low isoTOP-ABPP ratios could reflect the hyperreactivity of active cysteines, including nucleophilic and reductive catalysis, as well as those that undergo various forms of oxidative modifications.

The design of ABPs often relies on the covalent interactions between enzyme and substrate to irreversibly modify the conserved active-site nucleophiles (e.g. cysteine and lysine). However, many enzyme classes do not have active-site nucleophiles ready for ABP labelling. To target these enzyme families, researchers designed non-covalent ABPs that contain photocrosslinking groups to facilitate the enzyme binding. This strategy has proved to be successful in the design of ABPs for several enzyme families, such as metalloproteases [113], kinase [114], and also non-enzyme proteins. HDACs are known to play a critical role in cancer development, where the increased HDAC activity has generally been associated with transcriptional repression. However, the study of HDAC activity is hampered by the presence of endogenous activating protein complexes that are disrupted upon cell lysis. To address this problem, Salisbury and Cravatt developed an active-site-directed chemical probe to profile HDAC activities under living conditions [115]. They converted suberoylanilide hydroxamic acid (SAHA), a known reversible inhibitor for HDACs, into an irreversible ABP (SAHA-BPyne) by installation of an alkyne handle as well as a benzophenone photo affinity group which can promote covalent labeling of proximal proteins via photolysis (Fig. 10). Using this probe to profile the activity of HDACs in cancer cell proteome, the authors identified both class I and

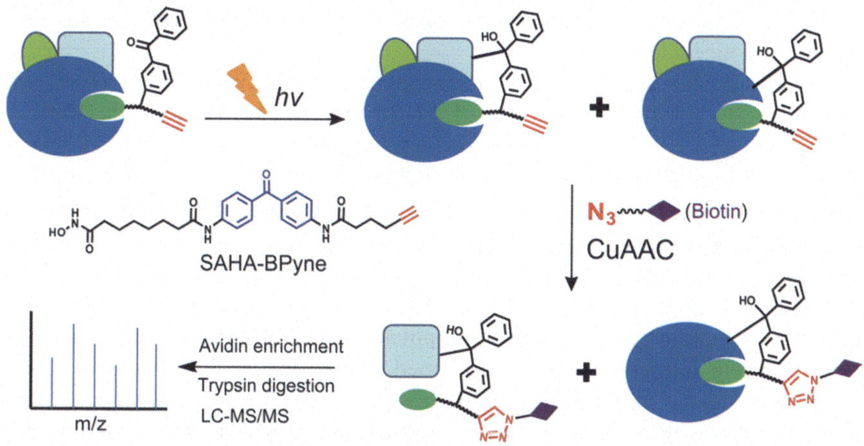

Fig. 10 Schematic representation for the photo-affinity-based ABPP strategy. The structure of HDAC photo-crosslinking probe (SAHA-BPyne) is shown in the middle [115]. *ABPP* activity-based protein profiling, *HDAC* histone deacetylase, *SAHA* suberoylanilide hydroxamic acid, *CuAAC* copper(I)-catalyzed azide–alkyne cycloaddition, *LC–MS/MS* liquid chromatography–tandem mass spectrometry

class II HDACs. Interestingly, many HDAC-associated proteins were also enriched by SAHA-BPyne, indicating that these HDAC-associated proteins were in close proximity to HDAC activity sites, where they may play certain functional roles such as regulating substrate recognition and activity.

Photo-affinity ABPs have also been applied to identify and characterize non-enzymatic activities of cognate receptors, such as cancer-cell marker galectin-3 [116] and nicotinic acetylcholine receptor (nAChR) [117]. In addition, such probes have also been utilized for target identification during drug discovery. A prominent example is the target identification for modulators of γ-secretase (GSMs) [118, 119], a critical target for the treatment of Alzheimer's disease. Labeling by photo-reactive GSMs revealed that multiple binding sites existed in the γ-secretase complex, each of which contributed to different modes of regulation on the activity of the complex. This elegant finding may help better understand the mechanism of γ-secretase and develop new active drugs towards Alzheimer's disease.

4 Ruthenium-Catalyzed Azide–Alkyne Cycloaddition (RuAAC)

To date, copper ions still remain the choice of metal catalyst for the azide–alkyne cycloaddition reaction, which generates 1,4-disubstituted 1,2,3-triazoles as the final product. Chemists have also studied the feasibility of other metals in catalyzing the azide–alkyne cycloaddition. For example, Hein and Fokin surveyed the complexes of all of the first-row transition elements, as well as complexes of Pd(0/II), Pt(II), Au(I/III), and Hg(II) ions, among others. However, none of these metal ions were found to produce triazoles in synthetically useful yields, and the effects of these metal ions on the rate and regioselectivity of the cycloaddition reaction was only marginally noticeable [32].

In 2005, Zhang and coworkers reported the Ru-catalyzed formation of the complementary 1,5-disubstituted triazoles [120]. Other than CuAAC, the RuAAC reaction can proceed between the azide and either terminal or internal alkyne moieties to form the cycloaddition product (Fig. 11a). The proposed mechanism is shown in Fig. 11b [121]. In the first step, both the azide and alkyne group are coordinated to the Ru(II) center by replacing two previous ligands. The azide group binds through the nitrogen atom proximal to the carbon atom. Notably, instead of a metal-acetylide intermediate, the alkyne group is coordinated to the Ru(II) center through its π-bond, which explains the applicability of RuAAC to internal alkynes. A ruthenacycle intermediate is then formed which directs the formation of the coordinated triazole. In the last step, the 1,5-disubstituted triazole is released from the Ru complex through a reductive elimination; however, this reaction only proceeds in organic solvents and the usage of inert gas atmosphere is often required for clean and smooth conversions. In addition, the RuAAC may have side reactions with carboxyl or amine groups. All these shortages limit their applications in biological systems. Nonetheless, the RuAAC reaction still has applications in

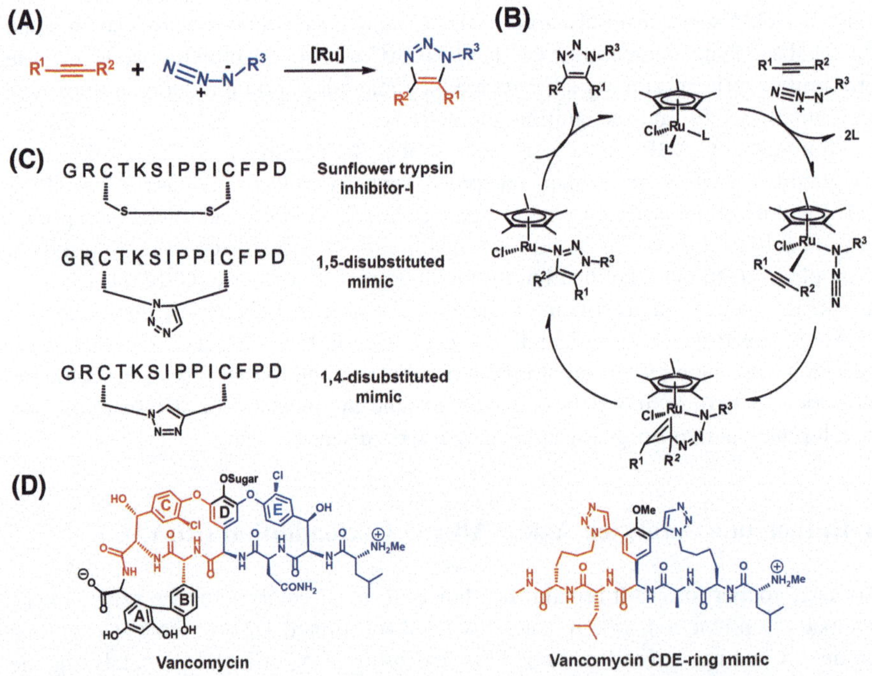

Fig. 11 Ru-catalyzed azide-alkyne cycloaddition reaction (RuAAC) **a** RuAAC reaction. **b** Proposed mechanism for RuAAC reaction. **c** 'Triazole bridge' generated by CuAAC or RuAAC in peptide mimics. d Structure of vancomycin and vancomycin CDE-ring mimic. The CD-ring of vancomycin is colored in *red* and the DE-ring is colored in *blue*. The CDE ring is colored in both *blue* and *red* [120–124]. *RuAAC* ruthenium-catalyzed azide–alkyne cycloaddition reaction, *CuAAC* copper(I)-catalyzed azide–alkyne cycloaddition

peptide mimics and drug development as a complementary approach to the CuAAC reaction [37].

RuAAC has been applied as the replacement of disulfide-bond through the formation of 'triazole bridge' in peptide mimics. Disulfide bond is essential in maintaining the conformation and functionality of bridged peptides. To increase the redox stability of bridged peptides, people have developed many strategies to mimic the disulfide bonds, among which 'triazole bridge' emerged as a valuable approach [123]. In complementary with the most commonly used 1,4-disubstituted 'triazole bridge' generated from the CuAAC reaction, the 1,5-disubstituted 'triazole bridge' generated from the RuAAC reaction expanded architecture of the peptide mimics. In 2010, the Empting et al. group used RuAAC to generate a 1,5-disubstituted 'triazole bridge' in a monocyclic variant of the sunflower trypsin inhibitor-I [122]. The resulting peptide mimic retained nearly full biological activity. In contrast, the peptide mimic generated by 1,4-disubstituted 'triazole bridge' with the CuAAC reaction retained only partial biological activity (Fig. 11c).

Recently, Zhang et al. applied RuAAC reaction to synthesize a 1,5-triazole bridged bicyclic structure as a mimic of the CDE-ring of vancomycin (Fig. 11d), which is a highly important glycopeptide antibiotic for treating bacterial infections [124]. By using molecular modelling, the authors found that 1,5-triazole bridged DE-ring mimic has the same ring size as the natural DE-ring and showed high structural resemblance to vancomycin (Fig. 11d).

In addition to the RuAAC reaction, McNulty and coworkers recently developed an Ag(I) complex in which Ag(I) ligated to a 2-diphenylphosphino-*N*,*N*,-diisopropylcarboxamide ligand that can catalyze the cycloaddition of azides onto terminal alkynes at room temperature to yield the 1,4-substitued triazoles, so called the AgAAC reaction [125]. The authors found that this reaction proceeded via Ag acetylide formation, and further activation of the acetylide–azide intermediate by the Ag(I) complex was necessary for cycloaddition. However, the concentration of the Ag(I) complex required for this reaction is as high as 20 mol %. In a later work, the same group optimized the ligand in the Ag(I) complex, and discovered a robust Ag(I) catalyst which was effective at loadings as low as 0.5 mol % [126]. More recently, Gao and coworkers developed a Ag-catalyzed cycloaddition of terminal alkynes with isocyanides by using Ag_2CO_3 as the catalyst. The authors demonstrated the application of this reaction in constructing substituted pyrroles, which proved to be extremely simple and efficient [127].

In addition to the Ru- and Ag-catalyzed cycloaddition reaction, Horneff and coworkers developed an Rh-catalyzed transannulation reaction of 1,2,3-triazoles with nitriles (Fig. 12) [128]. The CuAAC product *N*-sulfonyl 1,2,3-triazoles remains reactive and can react with nitriles to form the imidazole products through an Rh-

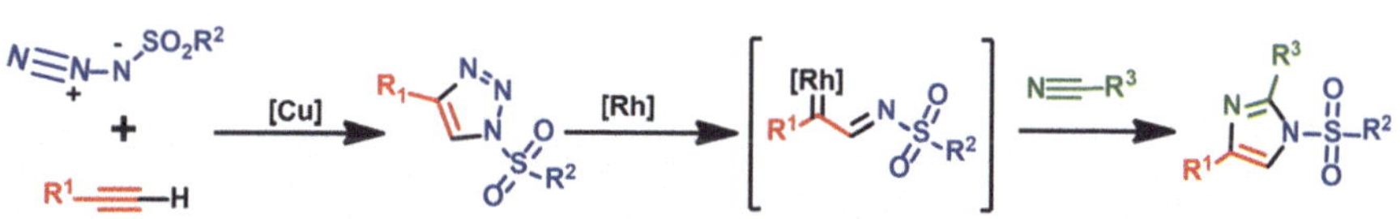

Fig. 12 Rh-catalyzed transannulation of 1,2,3-triazoles with nitriles [128]

catalyzed reaction. In the first step, Rh(II) complexes catalyze the ring opening of *N*-sulfonyl 1,2,3-triazoles to generate rhiminocarbenoids, which then react with the nitriles through the cycloaddition reaction. This reaction provides a concept that a second round of modification can be achieved in situ after the modification of target via CuAAC reaction. However, similar to RuAAC, this reaction also only proceeds in organic solvent, which limits its applications in biomolecule labeling. Nevertheless, this transannulation reaction has now been expanded to react with vinyl ethers and applied to the synthesis of some natural drug candidates [129].

Until now, all cycloaddition reactions catalyzed by transition metals other than Cu are limited to organic solvents. We expect such cycloaddition reactions to be optimized to proceed in aqueous solution at room temperature in the future, which would significantly expand their applications in biological systems.

5 Conclusions

The use of transition-metal catalysts to promote cycloaddition reactions has been an important strategy for selective construction of chemical bonds. Two key features for these transition-metal catalysts are their ability to (i) induce otherwise impossible reactions to occur (or reactions with a very low reaction rate); and (ii) alter the stereoselectivity of these reactions. Beyond carrying out these reactions within flasks, such transition-metal-catalyzed cycloaddition reactions have been extended to biomolecules within a biological context. As a hallmark, the CuAAC reaction has been widely adopted for labeling biomolecules in living systems. Recent developments of the ligand-assisted CuAAC reaction has circumvented the cytotoxicity of Cu(I) ions, with more progresses being made to label intracellular proteins inside *E. coli* cells. Meanwhile, other biomolecules, including glycans, lipids and nucleic acids, have all been successfully labeled, which greatly broadened the applicability of the CuAAC reaction. Additionally, other transition-metal-catalyzed cycloaddition reactions have been applied in synthetic chemistry, with their applications in biological systems still in their infancy. Ru-catalyzed azide–alkyne cycloaddition has broad scope and functional group compatibility but it is sensitive to the solvents. Rh-catalyzed transannulation of 1,2,3-triazoles with nitriles is an interesting process that provides a second-round modification after the CuAAC reaction. However, the requirement of a high temperature for this reaction has hindered its usage on biomolecules at the current stage. Future efforts will be made towards optimizing these reactions to meet the biocompatible criteria for living systems as well as exploring additional transition metals to catalyze such bioorthogonal cycloaddition reactions.

References

1. Wang Q, Chan TR, Hilgraf R, Fokin VV, Sharpless KB, Finn MG (2003) Bioconjugation by copper(I)-catalyzed azide–alkyne [3 + 2] cycloaddition. J Am Chem Soc 125:3192–3193

2. Fokin VV (2007) Click imaging of biochemical processes in living systems. ACS Chem Biol 2:775–778
3. Kolb HC, Sharpless KB (2003) The growing impact of click chemistry on drug discovery. Drug Discov Today 8:1128–1137
4. Zheng T, Rouhanifard SH, Jalloh AS, Wu P (2012) Click triazoles for bioconjugation. Top Heterocycl Chem 28:163–183
5. Gupta SS, Kuzelka J, Singh P, Lewis WG, Manchester M, Finn MG (2005) Accelerated bioorthogonal conjugation: a practical method for the ligation of diverse functional molecules to a polyvalent virus scaffold. Bioconjug Chem 16:1572–1579
6. Song W, Wang Y, Yu Z, Vera CIR, Qu J, Lin Q (2010) A metabolic alkene reporter for spatiotemporally controlled imaging of newly synthesized proteins in mammalian cells. ACS Chem Biol 5:875–885
7. Wang J, Zhang W, Song W, Wang Y, Yu Z, Li J, Wu M, Wang L, Zang J, Lin Q (2010) A biosynthetic route to photoclick chemistry on proteins. J Am Chem Soc 132:14812–14818
8. Agard NJ, Prescher JA, Bertozzi CR (2004) A strain-promoted [3 + 2] azide–alkyne cycloaddition for covalent modification of biomolecules in living systems. J Am Chem Soc 126:15046–15047
9. Blackman ML, Royzen M, Fox JM (2008) Tetrazine ligation: fast bioconjugation based on inverse-electron-demand Diels–Alder reactivity. J Am Chem Soc 130:13518–13519
10. Devaraj NK, Weissleder R, Hilderbrand SA (2008) Tetrazine-based cycloadditions: application to pretargeted live cell imaging. Bioconjug Chem 19:2297–2299
11. Prescher JA, Bertozzi CR (2005) Chemistry in living systems. Nat Chem Biol 1:13–21
12. Lim RKV, Lin Q (2010) Bioorthogonal chemistry: recent progress and future directions. Chem Commun 46:1589–1600
13. Grammel M, Hang HC (2013) Chemical reporters for biological discovery. Nat Chem Biol 9:475–484
14. Beatty KE, Fisk JD, Smart BP, Lu YY, Szychowski J, Hangauer MJ, Baskin JM, Bertozzi CR, Tirrell DA (2010) Live-cell imaging of cellular proteins by a strain-promoted azide–alkyne cycloaddition. Chembiochem 11:2092–2095
15. Lewis WG, Green LG, Grynszpan F, Radić Z, Carlier PR, Taylor P, Finn MG, Sharpless KB (2002) Click chemistry in situ: acetylcholinesterase as a reaction vessel for the selective assembly of a femtomolar inhibitor from an array of building blocks. Angew Chem Int Ed 41:1053–1057
16. Lewis WG, Magallon FG, Fokin VV, Finn MG (2004) Discovery and characterization of catalysts for azide–alkyne cycloaddition by fluorescence quenching. J Am Chem Soc 126:9152–9153
17. Presolski SI, Hong V, Cho S-H, Finn MG (2010) Tailored ligand acceleration of the cu-catalyzed azide–alkyne cycloaddition reaction: practical and mechanistic implications. J Am Chem Soc 132:14570–14576
18. Uttamapinant C, Tangpeerachaikul A, Grecian S, Clarke S, Singh U, Slade P, Gee KR, Ting AY (2012) Fast, cell-compatible click chemistry with copper-chelating azides for biomolecular labeling. Angew Chem Int Ed 51:5852–5856
19. Codelli JA, Baskin JM, Agard NJ, Bertozzi CR (2008) Second-generation difluorinated cyclooctynes for copper-free click chemistry. J Am Chem Soc 130:11486–11493
20. Ning X, Guo J, Wolfert MA, Boons G-J (2008) Visualizing metabolically labeled glycoconjugates of living cells by copper-free and fast huisgen cycloadditions. Angew Chem Int Ed 47:2253–2255
21. Baskin JM, Bertozzi CR (2007) Bioorthogonal click chemistry: covalent labeling in living systems. QSAR Comb Sci 26:1211–1219
22. Plass T, Milles S, Koehler C, Schultz C, Lemke EA (2011) Genetically encoded copper-free click chemistry. Angew Chem Int Ed 50:3878–3881
23. Lang K, Davis L, Torres-Kolbus J, Chou C, Deiters A, Chin JW (2012) Genetically encoded norbornene directs site-specific cellular protein labelling via a rapid bioorthogonal reaction. Nat Chem 4:298–304
24. Plass T, Milles S, Koehler C, Szymański J, Mueller R, Wießler M, Schultz C, Lemke EA (2012) Amino acids for Diels–Alder reactions in living cells. Angew Chem Int Ed 51:4166–4170
25. Song W, Wang Y, Qu J, Lin Q (2008) Selective functionalization of a genetically encoded alkene-containing protein via "photoclick chemistry" in bacterial cells. J Am Chem Soc 130:9654–9655
26. Song W, Wang Y, Qu J, Madden MM, Lin Q (2008) A photoinducible 1,3-dipolar cycloaddition reaction for rapid, selective modification of tetrazole-containing proteins. Angew Chem Int Ed 47:2832–2835

27. Yu Z, Pan Y, Wang Z, Wang J, Lin Q (2012) Genetically encoded cyclopropene directs rapid, photoclick-chemistry-mediated protein labeling in mammalian cells. Angew Chem Int Ed 51:10600–10604
28. An P, Yu Z, Lin Q (2013) Design of oligothiophene-based tetrazoles for laser-triggered photoclick chemistry in living cells. Chem Commun 49:9920–9922
29. Best MD (2009) Click chemistry and bioorthogonal reactions: unprecedented selectivity in the labeling of biological molecules. Biochemistry 48:6571–6584
30. El-Sagheer AH, Brown T (2010) Click chemistry with DNA. Chem Soc Rev 39:1388–1405
31. Yang M, Li J, Chen PR (2014) Transition metal-mediated bioorthogonal protein chemistry in living cells. Chem Soc Rev 43:6511–6526
32. Hein JE, Fokin VV (2010) Copper-catalyzed azide–alkyne cycloaddition (CuAAC) and beyond: new reactivity of copper(I) acetylides. Chem Soc Rev 39:1302–1315
33. Huisgen R (1963) 1,3-Dipolar cycloadditions. past and future. Angew Chem Int Ed 2:565–598
34. Rostovtsev VV, Green LG, Fokin VV, Sharpless KB (2002) A stepwise huisgen cycloaddition process: copper(I)-catalyzed regioselective "ligation" of azides and terminal alkynes. Angew Chem Int Ed 41:2596–2599
35. Tornøe CW, Christensen C, Meldal M (2002) Peptidotriazoles on solid phase: [1 − 3]-triazoles by regiospecific copper(I)-catalyzed 1,3-dipolar cycloadditions of terminal alkynes to azides. J Org Chem 67:3057–3064
36. Kolb HC, Finn MG, Sharpless KB (2001) Click chemistry: diverse chemical function from a few good reactions. Angew Chem Int Ed 40:2004–2021
37. Pasini D (2013) The click reaction as an efficient tool for the construction of macrocyclic structures. Molecules 18:9512–9530
38. Liang L, Astruc D (2011) The copper(I)-catalyzed alkyne-azide cycloaddition (CuAAC) "click" reaction and its applications. An overview. Coord Chem Rev 255:2933–2945
39. Hein C, Liu X-M, Wang D (2008) Click chemistry, a powerful tool for pharmaceutical sciences. Pharm Res 25:2216–2230
40. Himo F, Lovell T, Hilgraf R, Rostovtsev VV, Noodleman L, Sharpless KB, Fokin VV (2005) Copper(I)-catalyzed synthesis of azoles. DFT study predicts unprecedented reactivity and intermediates. J Am Chem Soc 127:210–216
41. Worrell BT, Malik JA, Fokin VV (2013) Direct evidence of a dinuclear copper intermediate in Cu(I)-catalyzed azide–alkyne cycloadditions. Science 340:457–460
42. Meldal M, Tornøe CW (2008) Cu-catalyzed azide–alkyne cycloaddition. Chem Rev 108:2952–3015
43. Chan TR, Hilgraf R, Sharpless KB, Fokin VV (2004) Polytriazoles as copper(I)-stabilizing ligands in catalysis. Org Lett 6:2853–2855
44. Rodionov VO, Presolski SI, Díaz Díaz D, Fokin VV, Finn MG (2007) Ligand-accelerated Cu-catalyzed azide–alkyne cycloaddition: a mechanistic report. J Am Chem Soc 129:12705–12712
45. Schoffelen S, Lambermon MHL, Eldijk MBV, Hest JCMV (2008) Site-specific modification of *Candida antarctica* lipase B via residue-specific incorporation of a non-canonical amino acid. Bioconjug Chem 19:1127–1131
46. Link AJ, Tirrell DA (2003) Cell surface labeling of escherichia coli via copper(I)-Catalyzed [3 + 2] cycloaddition. J Am Chem Soc 125:11164–11165
47. Hong V, Steinmetz NF, Manchester M, Finn MG (2010) Labeling live cells by copper-catalyzed alkyne–azide click chemistry. Bioconjug Chem 21:1912–1916
48. Besanceney-Webler C, Jiang H, Zheng T, Feng L, Soriano del Amo D, Wang W, Klivansky LM, Florence L, Liu Y, Wu P (2011) Increasing the efficacy of bioorthogonal click reactions for bioconjugation: a comparative study. Angew Chem Int Ed 50:8051–8056
49. Yang M, Song Y, Zhang M, Lin S, Hao Z, Liang Y, Zhang D, Chen PR (2012) Converting a solvatochromic fluorophore into a protein-based pH indicator for extreme acidity. Angew Chem Int Ed 51:7674–7679
50. Wang W, Hong S, Tran A, Jiang H, Triano R, Liu Y, Chen X, Wu P (2011) Sulfated ligands for the copper(I)-catalyzed azide–alkyne cycloaddition. Chem Asian J 6:2796–2802
51. Yang M, Jalloh AS, Wei W, Zhao J, Wu P, Chen PR (2014) Biocompatible click chemistry enabled compartment-specific pH measurement inside *E. coli*. Nat Commun 5:4981
52. Soriano del Amo D, Wang W, Jiang H, Besanceney C, Yan AC, Levy M, Liu Y, Marlow FL, Wu P (2010) Biocompatible copper(I) catalysts for in vivo imaging of glycans. J Am Chem Soc 132:16893–16899

53. Kennedy DC, McKay CS, Legault MCB, Danielson DC, Blake JA, Pegoraro AF, Stolow A, Mester Z, Pezacki JP (2011) Cellular consequences of copper complexes used to catalyze bioorthogonal click reactions. J Am Chem Soc 133:17993–18001
54. Link AJ, Vink MKS, Tirrell DA (2004) Presentation and detection of azide functionality in bacterial cell surface proteins. J Am Chem Soc 126:10598–10602
55. Hong V, Presolski SI, Ma C, Finn MG (2009) Analysis and optimization of copper-catalyzed azide–alkyne cycloaddition for bioconjugation. Angew Chem Int Ed 48:9879–9883
56. Brotherton WS, Michaels HA, Simmons JT, Clark RJ, Dalal NS, Zhu L (2009) Apparent copper(II)-accelerated azide–alkyne cycloaddition. Org Lett 11:4954–4957
57. Kuang G-C, Michaels HA, Simmons JT, Clark RJ, Zhu L (2010) Chelation-assisted, copper(II)-acetate-accelerated azide–alkyne cycloaddition. J Org Chem 75:6540–6548
58. Jiang H, Zheng T, Lopez-Aguilar A, Feng L, Kopp F, Marlow FL, Wu P (2014) Monitoring dynamic glycosylation in vivo using super-sensitive click chemistry. Bioconjug Chem 25:698–706
59. Sletten EM, Bertozzi CR (2009) Bioorthogonal chemistry: fishing for selectivity in a sea of functionality. Angew Chem Int Ed 48:6974–6998
60. Patterson DM, Nazarova LA, Prescher JA (2014) Finding the right (bioorthogonal) chemistry. ACS Chem Biol 9:592–605
61. Zheng M, Zheng L, Zhang P, Li J, Zhang Y (2015) Development of bioorthogonal reactions and their applications in bioconjugation. Molecules 20:3190
62. Thirumurugan P, Matosiuk D, Jozwiak K (2013) Click chemistry for drug development and diverse chemical-biology applications. Chem Rev 113:4905–4979
63. Peng T, Yuan X, Hang HC (2014) Turning the spotlight on protein–lipid interactions in cells. Curr Opin Chem Biol 21:144–153
64. Martell J, Weerapana E (2014) Applications of copper-catalyzed click chemistry in activity-based protein profiling. Molecules 19:1378–1393
65. El-Sagheer AH, Brown T (2012) Click nucleic acid ligation: applications in biology and nanotechnology. Acc Chem Res 45:1258–1267
66. Nischan N, Hackenberger CPR (2014) Site-specific PEGylation of proteins: recent developments. J Org Chem 79:10727–10733
67. Jing C, Cornish VW (2011) Chemical tags for labeling proteins inside living cells. Acc Chem Res 44:784–792
68. Lang K, Chin JW (2014) Bioorthogonal reactions for labeling proteins. ACS Chem Biol 9:16–20
69. Johnson JA, Lu YY, Van Deventer JA, Tirrell DA (2010) Residue-specific incorporation of non-canonical amino acids into proteins: recent developments and applications. Curr Opin Chem Biol 14:774–780
70. Liu CC, Schultz PG (2010) Adding new chemistries to the genetic code. Annu Rev Biochem 79:413–444
71. Hinner MJ, Johnsson K (2010) How to obtain labeled proteins and what to do with them. Curr Opin Biotechnol 21:766–776
72. Uttamapinant C, White KA, Baruah H, Thompson S, Fernández-Suárez M, Puthenveetil S, Ting AY (2010) A fluorophore ligase for site-specific protein labeling inside living cells. Proc Natl Acad Sci USA 107:10914–10919
73. Chin JW (2014) Expanding and reprogramming the genetic code of cells and animals. Annu Rev Biochem 83:379–408
74. Wang Q, Parrish AR, Wang L (2009) Expanding the genetic code for biological studies. Chem Biol 16:323–336
75. Tyagi S, Lemke EA (2013) Chapter 9—genetically encoded click chemistry for single-molecule FRET of proteins. In: Conn Pm (ed) vol 113, Academic Press, pp 169–187
76. Lin S, Yan H, Li L, Yang M, Peng B, Chen S, Li W, Chen PR (2013) Site-specific engineering of chemical functionalities on the surface of live hepatitis D virus. Angew Chem Int Ed 52:13970–13974
77. Beatty KE, Xie F, Wang Q, Tirrell DA (2005) Selective dye-labeling of newly synthesized proteins in bacterial cells. J Am Chem Soc 127:14150–14151
78. Beatty KE, Liu JC, Xie F, Dieterich DC, Schuman EM, Wang Q, Tirrell DA (2006) Fluorescence visualization of newly synthesized proteins in mammalian cells. Angew Chem Int Ed 118:7524–7527
79. Kang H, Schuman EM (1996) A requirement for local protein synthesis in neurotrophin-induced hippocampal synaptic plasticity. Science 273:1402–1406

80. Martin KC, Casadio A, Zhu H, Yaping E, Rose JC, Chen M, Bailey CH, Kandel ER (1997) Synapse-specific, long-term facilitation of aplysia sensory to motor synapses: a function for local protein synthesis in memory storage. Cell 91:927–938
81. Dieterich DC, Hodas JJL, Gouzer G, Shadrin IY, Ngo JT, Triller A, Tirrell DA, Schuman EM (2010) In situ visualization and dynamics of newly synthesized proteins in rat hippocampal neurons. Nat Neurosci 13:897–905
82. Lewin GR, Barde Y-A (1996) Physiology of the Neurotrophins. Annu Rev Neurosci 19:289–317
83. Schwarz F, Aebi M (2011) Mechanisms and principles of N-linked protein glycosylation. Curr Opin Struct Biol 21:576–582
84. Hang HC, Bertozzi CR (2005) The chemistry and biology of mucin-type O-linked glycosylation. Bioorg Med Chem 13:5021–5034
85. Chen X, Varki A (2010) Advances in the biology and chemistry of sialic acids. ACS Chem Biol 5:163–176
86. Dube DH, Bertozzi CR (2003) Metabolic oligosaccharide engineering as a tool for glycobiology. Curr Opin Chem Biol 7:616–625
87. Jiang H, English BP, Hazan RB, Wu P, Ovryn B (2015) Tracking surface glycans on live cancer cells with single-molecule sensitivity. Angew Chem Int Ed 54:1765–1769
88. Dube DH, Bertozzi CR (2005) Glycans in cancer and inflammation [mdash] potential for therapeutics and diagnostics. Nat Rev Drug Discov 4:477–488
89. Kannagi R, Izawa M, Koike T, Miyazaki K, Kimura N (2004) Carbohydrate-mediated cell adhesion in cancer metastasis and angiogenesis. Cancer Sci 95:377–384
90. Liu Y-C, Yen H-Y, Chen C-Y, Chen C-H, Cheng P-F, Juan Y-H, Chen C-H, Khoo K-H, Yu C-J, Yang P-C, Hsu T-L, Wong C-H (2011) Sialylation and fucosylation of epidermal growth factor receptor suppress its dimerization and activation in lung cancer cells. Proc Natl Acad Sci USA 108:11332–11337
91. Du J, Hong S, Dong L, Cheng B, Lin L, Zhao B, Chen Y-G, Chen X (2015) Dynamic sialylation in transforming growth factor-β (TGF-β)-induced epithelial to mesenchymal transition. J Biol Chem 290:12000–12013
92. Woo CM, Iavarone AT, Spiciarich DR, Palaniappan KK, Bertozzi CR (2015) Isotope-targeted glycoproteomics (IsoTaG): a mass-independent platform for intact N- and O-glycopeptide discovery and analysis. Nat Meth 12:561–567
93. van Meer G, Voelker DR, Feigenson GW (2008) Membrane lipids: where they are and how they behave. Nat Rev Mol Cell Biol 9:112–124
94. Resh MD (2006) Trafficking and signaling by fatty-acylated and prenylated proteins. Nat Chem Biol 2:584–590
95. Hang HC, Wilson JP, Charron G (2011) Bioorthogonal chemical reporters for analyzing protein lipidation and lipid trafficking. Acc Chem Res 44:699–708
96. Jao C, Roth M, Welti R, Salic A (2009) Metabolic labeling and direct imaging of choline phospholipids in vivo. Proc Natl Acad Sci USA 106:15332–15337
97. Yount JS, Moltedo B, Yang Y-Y, Charron G, Moran TM, López CB, Hang HC (2010) Palmitoylome profiling reveals S-palmitoylation–dependent antiviral activity of IFITM3. Nat Chem Biol 6:610–614
98. Peng T, Hang HC (2015) Bifunctional fatty acid chemical reporter for analyzing S-palmitoylated membrane protein-protein interactions in mammalian cells. J Am Chem Soc 137:556–559
99. Niphakis MJ, Lum KM, Cognetta Iii AB, Correia BE, Ichu T-A, Olucha J, Brown SJ, Kundu S, Piscitelli F, Rosen H, Cravatt BF (2015) A global map of lipid-binding proteins and their ligandability in cells. Cell 161:1668–1680
100. Haberkant P, Raijmakers R, Wildwater M, Sachsenheimer T, Brügger B, Maeda K, Houweling M, Gavin A-C, Schultz C, van Meer G, Heck AJR, Holthuis JCM (2013) In vivo profiling and visualization of cellular protein-lipid interactions using bifunctional fatty acids. Angew Chem Int Ed 52:4033–4038
101. Hulce JJ, Cognetta AB, Niphakis MJ, Tully SE, Cravatt BF (2013) Proteome-wide mapping of cholesterol-interacting proteins in mammalian cells. Nat. Meth 10:259–264
102. Salic A, Mitchison TJ (2008) A chemical method for fast and sensitive detection of DNA synthesis in vivo. Proc Natl Acad Sci USA 105:2415–2420
103. Jao CY, Salic A (2008) Exploring RNA transcription and turnover in vivo by using click chemistry. Proc Natl Acad Sci USA 105:15779–15784

104. Neef AB, Pernot L, Schreier VN, Scapozza L, Luedtke NW (2015) A bioorthogonal chemical reporter of viral infection. Angew Chem Int Ed 127:8022–8025
105. Cravatt BF, Sorensen EJ (2000) Chemical strategies for the global analysis of protein function. Curr Opin Chem Biol 4:663–668
106. Evans MJ, Cravatt BF (2006) Mechanism-based profiling of enzyme families. Chem Rev 106:3279–3301
107. Sadaghiani AM, Verhelst SHL, Bogyo M (2007) Tagging and detection strategies for activity-based proteomics. Curr Opin Chem Biol 11:20–28
108. Speers AE, Cravatt BF (2004) Profiling enzyme activities in vivo using click chemistry methods. Chem Biol 11:535–546
109. Cravatt BF, Wright AT, Kozarich JW (2008) Activity-based protein profiling: from enzyme chemistry to proteomic chemistry. Annu Rev Biochem 77:383–414
110. Fonović M, Bogyo M (2008) Activity-based probes as a tool for functional proteomic analysis of proteases. Expert Rev Proteomics 5:721–730
111. Speers AE, Cravatt BF (2005) A tandem orthogonal proteolysis strategy for high-content chemical proteomics. J Am Chem Soc 127:10018–10019
112. Weerapana E, Wang C, Simon GM, Richter F, Khare S, Dillon MBD, Bachovchin DA, Mowen K, Baker D, Cravatt BF (2010) Quantitative reactivity profiling predicts functional cysteines in proteomes. Nature 468:790–795
113. Sieber SA, Niessen S, Hoover HS, Cravatt BF (2006) Proteomic profiling of metalloprotease activities with cocktails of active-site probes. Nat Chem Biol 2:274–281
114. Kalesh KA, Sim DSB, Wang J, Liu K, Lin Q, Yao SQ (2010) Small molecule probes that target Abl kinase. Chem Commun 46:1118–1120
115. Salisbury CM, Cravatt BF (2007) Activity-based probes for proteomic profiling of histone deacetylase complexes. Proc Natl Acad Sci USA 104:1171–1176
116. van Scherpenzeel M, van der Pot M, Arnusch CJ, Liskamp RMJ, Pieters RJ (2007) Detection of galectin-3 by novel peptidic photoprobes. Bioorg Med Chem Lett 17:376–378
117. Tantama M, Lin W-C, Licht S (2008) An activity-based protein profiling probe for the nicotinic acetylcholine receptor. J Am Chem Soc 130:15766–15767
118. Crump CJ, am Ende CW, Eric Ballard T, Pozdnyakov N, Pettersson M, Chau D-M, Bales KR, Li Y-M, Johnson DS (2012) Development of clickable active site-directed photoaffinity probes for γ-secretase. Bioorg Med Chem Lett 22:2997–3000
119. Pozdnyakov N, Murrey HE, Crump CJ, Pettersson M, Ballard TE, am Ende CW, Ahn K, Li Y-M, Bales KR, Johnson DS (2013) γ-Secretase modulator (GSM) PHOTOAFFINITY probes reveal distinct allosteric binding sites on presenilin. J Biol Chem 288:9710–9720
120. Zhang L, Chen X, Xue P, Sun HHY, Williams ID, Sharpless KB, Fokin VV, Jia G (2005) Ruthenium-catalyzed cycloaddition of alkynes and organic azides. J Am Chem Soc 127:15998–15999
121. Boren BC, Narayan S, Rasmussen LK, Zhang L, Zhao H, Lin Z, Jia G, Fokin VV (2008) Ruthenium-catalyzed azide–alkyne cycloaddition: scope and mechanism. J Am Chem Soc 130:8923–8930
122. Empting M, Avrutina O, Meusinger R, Fabritz S, Reinwarth M, Biesalski M, Voigt S, Buntkowsky G, Kolmar H (2011) "Triazole bridge": disulfide-bond replacement by ruthenium-catalyzed formation of 1,5-disubstituted 1,2,3-triazoles. Angew Chem Int Ed 50:5207–5211
123. Roice M, Johannsen I, Meldal M (2004) High capacity poly(ethylene glycol) based amino polymers for peptide and organic synthesis. QSAR Comb Sci 23:662–673
124. Zhang J, Kemmink J, Rijkers DTS, Liskamp RMJ (2013) Synthesis of 1,5-triazole bridged vancomycin CDE-ring bicyclic mimics using RuAAC macrocyclization. Chem Commun 49:4498–4500
125. McNulty J, Keskar K, Vemula R (2011) The first well-defined silver(I)-complex-catalyzed cycloaddition of azides onto terminal alkynes at room temperature. Chem Eur J 17:14727–14730
126. McNulty J, Keskar K (2012) Discovery of a robust and efficient homogeneous silver(I) catalyst for the cycloaddition of azides onto terminal alkynes. Eur J Org Chem 2012:5462–5470
127. Gao M, He C, Chen H, Bai R, Cheng B, Lei A (2013) Synthesis of pyrroles by click reaction: silver-catalyzed cycloaddition of terminal alkynes with isocyanides. Angew Chem Int Ed 52:6958–6961
128. Horneff T, Chuprakov S, Chernyak N, Gevorgyan V, Fokin VV (2008) Rhodium-catalyzed transannulation of 1,2,3-triazoles with nitriles. J Am Chem Soc 130:14972–14974
129. Rajasekar S, Anbarasan P (2014) Rhodium-catalyzed transannulation of 1,2,3-triazoles to polysubstituted pyrroles. J Org Chem 79:8428–8434

Top Curr Chem (Z) (2016) 374:16
DOI 10.1007/s41061-016-0016-4

REVIEW

Strain-Promoted 1,3-Dipolar Cycloaddition of Cycloalkynes and Organic Azides

Jan Dommerholt[1] · Floris P. J. T. Rutjes[1] · Floris L. van Delft[2]

Received: 24 November 2015 / Accepted: 17 February 2016 / Published online: 22 March 2016

Abstract A nearly forgotten reaction discovered more than 60 years ago—the cycloaddition of a cyclic alkyne and an organic azide, leading to an aromatic triazole—enjoys a remarkable popularity. Originally discovered out of pure chemical curiosity, and dusted off early this century as an efficient and clean bioconjugation tool, the usefulness of cyclooctyne–azide cycloaddition is now adopted in a wide range of fields of chemical science and beyond. Its ease of operation, broad solvent compatibility, 100 % atom efficiency, and the high stability of the resulting triazole product, just to name a few aspects, have catapulted this so-called strain-promoted azide–alkyne cycloaddition (SPAAC) right into the top-shelf of the toolbox of chemical biologists, material scientists, biotechnologists, medicinal chemists, and more. In this chapter, a brief historic overview of cycloalkynes is provided first, along with the main synthetic strategies to prepare cycloalkynes and their chemical reactivities. Core aspects of the strain-promoted reaction of cycloalkynes with azides are covered, as well as tools to achieve further reaction acceleration by means of modulation of cycloalkyne structure, nature of azide, and choice of solvent.

Keywords Strain-promoted cycloaddition · Cyclooctyne · BCN · DIBAC · Azide

This article is part of the Topical Collection "Cycloadditions in Bioorthogonal Chemistry"; edited by Milan Vrabel, Thomas Carell

✉ Floris P. J. T. Rutjes
f.rutjes@science.ru.nl

✉ Floris L. van Delft
floris.vandelft@wur.nl

1 Radboud University, Nijmegen, The Netherlands

2 Wageningen University and Research Centre, Wageningen, The Netherlands

1 Introduction

The spontaneous reaction of cycloalkynes with an organic azide, in all its simplicity, is a fascinating organic chemical transformation. Simply by mixing and stirring, without the necessity of reagents, catalysts, or carefully controlled reaction conditions, a stable triazole product is formed by fast and selective cycloaddition of cycloalkyne with azide. As will become clear throughout this chapter, the latter reaction has now firmly established itself as a powerful and versatile chemical process with broad academic and commercial applications. Core to the chemistry lies a highly strained, medium-sized cyclic alkyne, most prominently a cyclooctyne. In this chapter, the synthesis and chemical reactivity of cyclic alkynes is broadly delineated, with particular emphasis on undoubtedly the most important of applications of cycloalkynes: cycloaddition with an organic azide, leading to the formation of a stable triazole.

2 The Fascinating Chemistry of Cycloalkynes

2.1 Conception of Cycloalkynes

In the second half of the previous century, interest emerged at several laboratories around the world to explore the synthesis and properties of medium-sized cycloalkynes. Pioneers in the field, Blomquist et al., at Cornell University (USA), convincingly demonstrated in 1951 that plain cyclononyne and cyclodecyne could be accessed by oxidative decomposition of the respective cycloalka-1,2-diones, and isolated in pure form by distillation [7]. Two years later, the same group also reported the successful preparation of the eight-membered ring acetylene [8], while similar explorations on cycloalkynes were reported by Prelog and colleagues at the ETH in Zurich (Switzerland) around the same time [44]. It must be noted that the preparation and isolation of cyclooctyne had been claimed by Domnin (USSR) some 15 years earlier [21], but the reported characterization data suggest that the compound was—at best—obtained as a mixture with the isomeric 1,2-cyclooctadiene and other unsaturated hydrocarbons. In 1961, it was Wittig at the University of Heidelberg (Germany) who was the first to demonstrate the formation of the yet-smaller five-, six-, and seven-membered cycloalkynes, as well as 1,2-dehydrobenzene, better known as benzyne [62].

The successful preparation of cycloalkynes also opened up the possibility to explore their unique chemical reactivity. In fact, the transient existence of the cycloalkyne species could initially only be indirectly corroborated by fast in situ trapping of the smaller-sized rings (seven carbons and below) before decomposition [31]. While not strictly applicable to cyclooctyne, which is the smallest cyclic alkyne that can be isolated and stored in pure form, Blomquist already noted that nevertheless careful exclusion of air was requisite to avoid rapid decomposition. More importantly, he was also the first to observe that "cyclooctyne reacts explosively when treated with phenyl azide, forming a viscous liquid product" [8]. This remark is in fact the first historic administration of a process that has now become known as strain-promoted azide–alkyne cycloaddition (SPAAC).

2.2 Synthetic Preparation of Cycloalkynes

A range of different synthetic procedures is known to obtain medium-sized cycloalkynes, as covered by several reviews [38]; Tochtermann [25]. In this paragraph, only the most relevant procedures will be covered.

The first synthetic reports on cycloalkynes involved a base-catalyzed oxidative decomposition of bis-hydrazones, readily prepared from the respective precursor 1,2-cycloalkadiones by condensation with hydrazine (Fig. 1, top). As oxidant, mercury oxide is most typically applied, however Ag_2O or $Pb(OAc)_4$ are also suitable. An analogous procedure (Fig. 1, bottom) employs the tosylate (Ts) derivative of hydrazine, which upon condensation with the diketone under reflux conditions forms the tosylated aminotriazole intermediate. Acidic removal of the tosyl group, followed by lead-mediated oxidation, also forms the desired cycloalkyne ring.

A direct procedure to obtain a cycloalkyne from a cycloalkanone involves the conversion into semicarbazone, followed by oxidation with selenium dioxide. The resulting intermediate 1,2,3-selenadiazole can be isolated in pure form and will eliminate, upon heating to 170–220 °C (or refluxing ethylene glycol), elemental selenium and nitrogen, with formation of the cycloalkyne.

The most often applied and most reliable procedure to obtain a cycloalkyne (Fig. 2, bottom), as originally developed by Brandsma in the Netherlands, [11], involves the stepwise double dehydrohalogenation of a vicinal dihalogenide (typically bromide), which can be readily obtained from a cyclic alkene upon treatment with elemental halogen. In this case, a first E2-elimination may take place under mildly basic conditions, forming an intermediate alkene, which in turn will undergo a second elimination upon treatment with a strong base like sodium amide or potassium *tert*-butoxide. More conveniently even, in many cases both

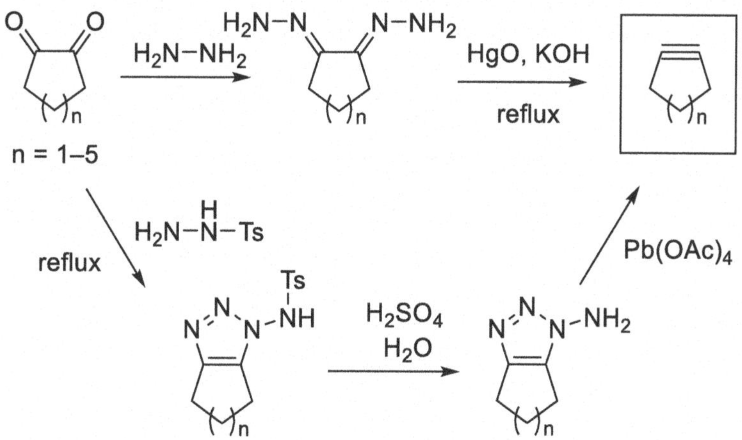

Fig. 1 Oxidative decomposition of cycloalka-1,2-dione dihydrazone, leading to medium-sized cycloalkynes (n = 1–5)

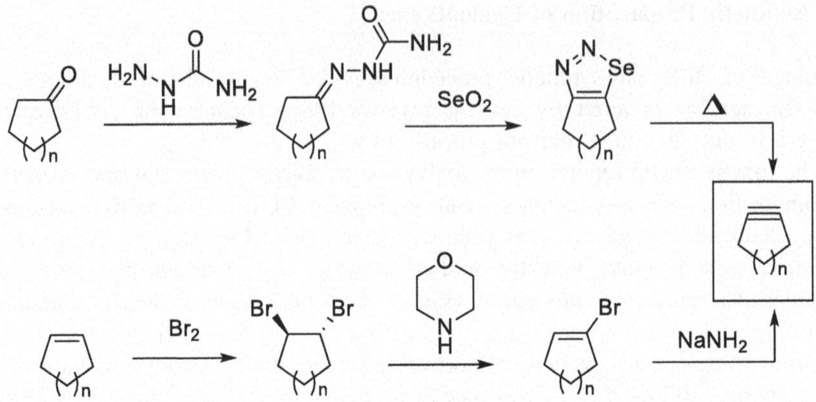

Fig. 2 Formation of cycloalkynes by fragmentation of 1,2,3-selenadiazole (*top*) or dehydrohalogenation (*bottom*)

eliminations can be induced in a one-step process upon treatment with a large excess of a strong base and/or heating.

Some more exotic synthetic processes for the synthesis of cycloalkynes, including reductive dehalogenation, photolytic elimination of bis-hydrazones or fragmentation of α,β-epoxyketones have also been reported, but are beyond the scope of this chapter.

2.3 Reactions of Cycloalkynes

The stability of cyclic alkynes rapidly decreases with ring size. In fact, stability is directly correlated with the C–C C–C bond angle, which, by virtue of the cyclic structure, cannot adopt the ideal 180° bond angle of sp-hybridized carbon atoms. Interestingly, cyclooctyne was identified as the smallest isolable cycloalkyne, although its acetylene bond angle of 163° still significantly deviates from linear. The experimentally determined ring strain of cyclooctyne is ~18 kcal/mol [58], compared to 12.1 kcal/mol for saturated cyclooctane [6]. Not surprisingly, ring-strain is accountable for the intrinsically low stability of medium-sized cyclic alkynes, with calculated ring strains of 25 kcal/mol and above [6], leading to fast degradation and/or polymerization of seven-membered and smaller cycloalkynes, and thwarting their isolation in pure form. Gratifyingly, the same ring strain also imparts a unique reactivity profile onto medium-sized cycloalkynes, which may be advantageously employed in many ways as described here. In fact, the first synthetic preparation of seven-, six-, and five-membered cycloalkynes by Wittig et al. in [62] could only be corroborated by in situ generation of the formed alkyne in the presence of a suitable 'alkynophile' for fast (4 + 2) cycloaddition [62]. In particular, 1,3-diphenylisobenzofuran (as illustrated for cycloheptyne in Fig. 3, left) was used, forming a stable oxanorbornene (4 + 2)-cycloadduct, suitable for isolation and characterization. An alternative quenching reagent for small-ring

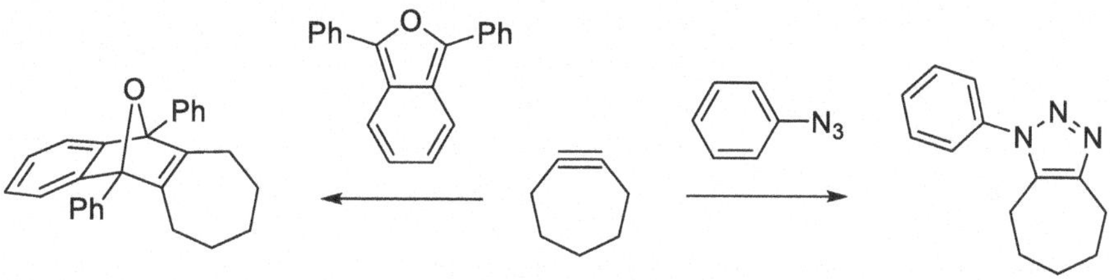

Fig. 3 Reaction of cycloheptyne with 1,3-diphenylisobenzofuran or phenyl azide, leading to bicyclic oxanorbornene (*left*) or phenyltriazole (*right*), respectively

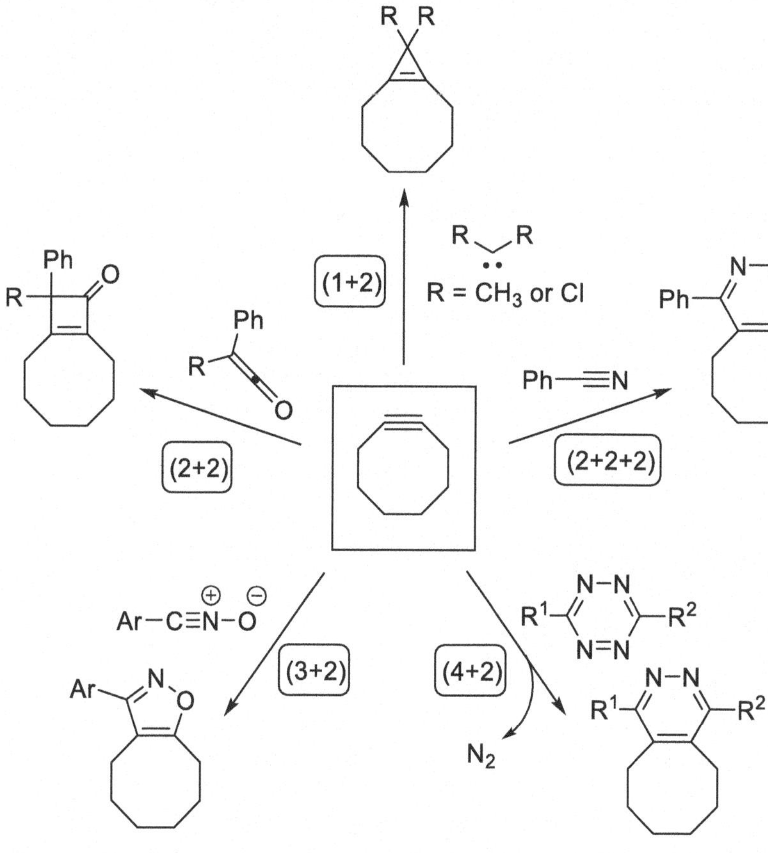

Fig. 4 Select examples of cycloaddition reactions of cyclooctyne

cycloalkynes is phenyl azide (Fig. 3, right), which will form a stable aromatic triazole by (3 + 2)-dipolar cycloaddition.

Since these first reports on the reactions of cycloalkynes with isobenzofuran and phenyl azide, a wide range of other alkynophile cycloaddition partners have become known, as conceptually illustrated with model compounds in reactions with cyclooctyne in Fig. 4. It becomes immediately clear that cycloalkynes can undergo a wide range of cycloaddition reactions, including (1 + 2), (2 + 2), (2 + 2 + 2), (3 + 2), and (4 + 2) cycloadditions. Besides these, more exotic transformations like (6 + 2) cycloadditions, hydrogen transfer reactions, radical additions, and reactions with metal salts or complexes, are also known. For further information on this topic, the reader is referred to several earlier reviews [25, 32].

2.4 Dipolar Cycloaddition of Cycloalkynes with Azides

The prototypical example of the reactivity of cycloalkynes in organic chemistry transformations, as already noted in the first report on the isolation of pure cyclooctyne [8], is the (3 + 2) dipolar cycloaddition with organic azides. A logical explanation for the fast and spontaneous reaction of cycloalkynes with (phenyl)azide therefore lies in the highly favorable enthalpic release of ring-strain, by going from a strained ring to a fused ring system with favorable bond angles for the sp^2-hybridized carbon atoms of the resulting triazole. It has been calculated [6] that the barrier of activation for (3 + 2) cycloaddition is directly correlated to strain energy of cycloalkynes. Houk et al. performed calculations on the transition state of the Huisgen 1,3-dipolar cycloadditions of phenyl azide with acetylene and cyclooctyne with density functional theory at the B3LYP level [22]. The low activation energy of the cyclooctyne cycloaddition ($\Delta E^{\ddagger} = 8.0$ kcal/mol) compared to the strain-free acetylene cycloaddition ($\Delta E^{\ddagger} = 16.2$ kcal/mol) was explained due to the decreased distortion energy for cyclooctyne to reach the requisite C–C–C bond angle of 158°–166° in the transition state versus the linear alkyne, i.e., deformation from 153° and 180°, respectively. Given the alkyne ring strain, the reaction of an organic azide with a cyclic alkyne, typically cyclooctyne, has become commonly known as strain-promoted azide–alkyne cycloaddition (SPAAC).

3 Speeding Up SPAAC

3.1 Copper-Free Click Reaction

With most of the activity around addition reactions of cyclic alkynes taking place in the last century, interest in this particular subclass of chemistry presumably would have slowly subsided if not for the clever insight by researchers at the University of California, Berkeley [1] that strain-promoted cycloaddition of cyclooctynes with azides is a highly versatile copper-free version of the popular click reaction [46, 57]. It was reasoned that remote attachment of a suitable functional handle to the cyclooctyne would enable the smooth conjugation to any organic azide, in any solvent of choice. In particular, it had become clear that the use of the copper-catalyzed azide–alkyne cycloaddition (CuAAC) was severely compromised in the context of biological matter, due to the toxicity of the inevitable copper(I)-species to living cells and organisms. In a seminal paper [1], it was shown that chemical functionalization of cyclooctyne with (+)-biotin enabled clean and selective visualization of living cells with azidosugars metabolically incorporated on the cell surface glycans, upon subjecting the cells to a SPAAC protocol with biotin-cyclooctyne followed by fluorescent labeling with a streptavidin-fluorophore. The broad impact of this seminal application of SPAAC, referred to in popular terms as "copper-free click reaction", can hardly be overestimated.

3.2 The Quest for More Reactive Cycloalkynes

While the first application of SPAAC rapidly found its way also outside the field of metabolic labeling, it also became quickly apparent that the relatively slow reaction kinetics required large excesses of reagents, long incubation times, and led to relatively little signal. In fact, visualization of metabolic labeling of azido-modified living cells was initially even less efficient than by Staudinger–Bertozzi ligation [49], a process known to suffer from poor reaction kinetics as well as oxygen sensitivity of the phosphine probe. As a result, a range of laboratories around the world, including ours, have embarked on the synthesis and evaluation of new cycloalkyne probes with the aim of lifting the reactivity without compromising on stability. A comprehensive overview of functionalized cyclooctynes that have been developed in the past decade, categorized by year of discovery, is provided in Fig. 5.

In general, two classes of cyclooctynes can be recognized: the earliest generation aliphatic cyclooctynes, and (di)benzoannulated cyclooctynes. The first dibenzoannulated cyclooctyne (DIBO) suitable for conjugation, developed by Ning et al. [40], has led the field to more reactive probes. It is commonly understood that the enhanced reactivity of (di)benzoannulated cyclooctynes is caused by the increase in ring strain imparted by the multiple sp^2-hybridized carbons. The latter phenomenon can also account for the reactivity order BARAC > DIBAC > DIBO, given that the number of sp^2-hybridized atoms in the ring decreases from 6 to 4, respectively. BARAC is in fact an interesting example of the fine balancing act between reactivity and stability that comes along with the development of cyclooctyne probes: BARAC displays a reaction rate constant of nearly 1 $mol^{-1}\ s^{-1}$ (see Table 1), but unfortunately is inherently unstable and rapidly decomposes [28]. Two cyclooctyne probes difluorobenzocycloocytne (DIFBO [55]) and 3,3,6,6-tetramethylthiaheptyne (TMTH [5]) are yet more reactive than BARAC but cannot be isolated in pure form before rapid decomposition takes place (not depicted in Fig. 5). Efforts from our own laboratory led to the development of DIBAC [16], a cyclooctyne that combined excellent stability with a reaction rate constant that is top among the family members. Interestingly, Kuzmin et al. [33] and Campbell-Verduijn et al. [13] rapidly followed with publications of the same molecule, albeit obtained by different synthetic paths and with different given names (ADIBO and aza-DBCO, respectively). In this chapter, the original term DIBAC will be consistently used to denote this azacyclooctyne structure, although DBCO is also often applied.

Besides reactivity, which is typically determined in organic (co)solvents like MeOH or MeCN, two other factors that qualify a given cyclooctyne, namely lipophilicity and size, are of high importance. For example, benzoannulation has shown to be beneficial for cyclooctyne reactivity but inevitably also leads to concomitant enhanced steric interactions and lipophilicity, which is typically not

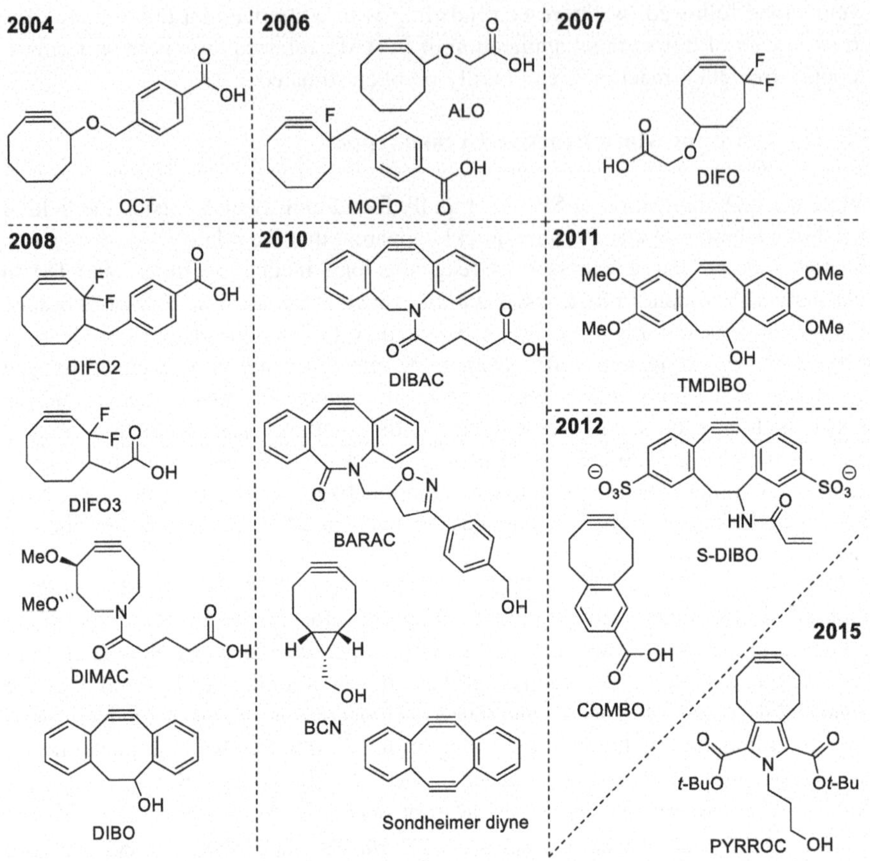

Fig. 5 Overview of functionalized cyclooctynes suitable for conjugation reactions, grouped by year of discovery. *OCT* cyclooctyne, *MOFO* monofluorinated cyclooctyne, *DIFO* difluorocyclooctyne, *DIMAC* dimethoxyazacyclooctyne, *DIBO* dibenzocyclooctyne, *DIBAC* dibenzoazacyclooctyne, *BARAC* biarylazacyclooctynone, *BCN* bicyclononyne, *TMDIBO* 2,3,6,7-tetramethoxy-DIBO, *S-DIBO* sulfonylated DIBO, *COMBO* carboxymethylmonobenzocyclooctyne, *PYRROC* pyrrolocyclooctyne

beneficial when SPAAC in water is envisaged. The first attempt to address the issue of lipophilicity was made by introduction of methoxy-groups on the cyclooctyne ring, as in DIMAC [54]. Although displaying excellent water solubility, reactivity of DIMAC was also severely compromised. Several more hydrophilic variants of DIBO have also been developed over the years, most prominently TMDIBO [56] and S-DIBO [23], both of which show much improved solubility in water, but the reaction rate constants remain rather low, as for the parent DIBO structure. The cyclooctynes COMBO [60] and PYRROC [24] have most recently been developed and are characterized by an intermediate, monobenzoannulated structure. Not unexpectedly, the reaction rate constants of COMBO and PYRROC are also lower than those for the analogous dibenzoannulated structures (Table 1). One notable exception to the rule that benzoannulation is a necessity to achieve acceptable reactivity is BCN, developed in our own laboratory [19]. For BCN, reactivity is induced

Table 1 Reaction rate constants (for BnN_3 or similar aliphatic azide) and synthetic accessibility of practically stable cyclooctynes

Entry	Cyclooctyne	k ($\times 10^{-3}$ M^{-1} s^{-1})	Solvent	#Steps	Yield (%)
1	OCT	2.4	A	4	52
2	DIMAC	3.0	A	11	5
3	MOFO	4.3	A	5	15
4	PYRROC	6.0	A	11	3
5	Sondheimer	8.8	C	4	41
6	DIFO2	42	A	8	27
7	DIFO3	52	A	10	21
8	DIFO	76	A	10	1
9	TMDIBO	94	C	7	57
10	S-DIBO	112	C	7	13
11	DIBO	120	B	5	10
12	BCN	140	C	4	15
13	COMBO	235	A	6	11
14	DIBAC	310	C	9	41
15	BARAC	960	A	6	18

Cyclooctynes are listed in order of reactivity. Solvent: A = CD_3CN, B = CD_3CN:D_2O (3:1), C = CD_3OD or CH_3OH

by ring fusion of cyclooctyne to cyclopropane, leading to the typical bicyclo[6.1.0]non-4-yne structure. Although less reactive than DIBAC, reaction rate constants for the *endo*-isomer are still among the highest in the pack (see Table 1), while the synthesis of BCN is exceptionally short and simple.

From Fig. 5, it also becomes apparent that while active development of cyclooctynes took place in the years 2008–2010, the intensity in the field has more or less subsided in the past years. One possible reason for this observation may be found in the fact that further boosting of the reactivity of cyclooctyne for azide is typically penalized by loss in stability, as earlier mentioned for BARAC, DIFBO, and TMTH. Another explanation lies in the current commercial availability of the cyclooctynes DIBO, DIBAC, and BCN, the three of which have dominated the field of SPAAC in the past years.

3.3 Influence of Azide Structure on Reaction Rate

While significant effort has been devoted over the years to the development of more reactive cyclooctynes, as delineated above, only scant investigations so far have focused on the increase of SPAAC rates by modulation of the complementary component, i.e., the azide. In fact, the vast majority of reported applications of SPAAC are based on reaction with simple aliphatic azides. As a logical consequence, reaction rate constants are also nearly always determined with an aliphatic azide (typically benzyl azide), but seldom with an aryl azide. One possible reason that aromatic azides are generally avoided for SPAAC may lie in a report

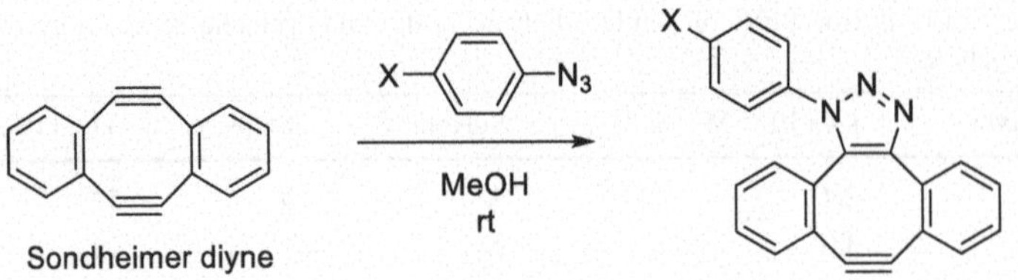

X	k_{rel}
H	1
OMe	3.8
CF_3	0.9

Fig. 6 Negligible influence of azidobenzene *para*-substitution on reaction rate with dibenzoannulated cyclooctyne

that *p*-azidophenylalanine shows sevenfold lower reactivity than that of the corresponding aliphatic azidomethyl analogue [66], at least in conjunction with DIBAC, a dibenzoannulated cyclooctyne. Furthermore, the observation that reaction rates of aromatic azides are hardly influenced by changing the electronic nature of substituents (as determined for *p*-methoxy and *p*-CF_3-phenyl azide, see Fig. 6) may have provided further ground to avoid aromatic azides for SPAAC [64]. These observations obviously support the notion that aryl azides, and in particular electron-poor azides (as in *p*-CF_3-PhN_3), are better avoided in case high SPAAC reaction rates are desirable. Other studies have reported a reactivity enhancement (up to 2.2× faster) upon the introduction of electron-withdrawing halogen substituents on DIFO [4], BARAC [28] and DIBAC [14, 18], all of which suggests that the SPAAC mechanism primarily proceeds via $HOMO_{azide}$–$LUMO_{cyclooctyne}$ interaction so that electron-rich azides are preferred. Interestingly, it has been known since the 1960s that electronegative substituents on aryl azides accelerate reaction rates with strained alkenes, for example a fourfold reaction rate enhancement of *p*-nitro substitution of phenyl azide with norbornene [50]. Even more markedly, picryl azide was found to react with norbornene almost 1000 times faster than phenyl azide [2].

Based on these observations, we recently concluded that the apparent slower reaction of electron-poor azides in SPAAC only holds for benzoannulated cyclooctynes (with low-lying LUMO), while in combination with more electron-rich cyclooctynes (like BCN), azides can react by a second, inverse electron-demand mechanism (i.e., SPAAC) as well [20]. As a result, it was first found that strain-promoted cycloaddition of aromatic azides with BCN is nearly eight times faster than with DIBAC, which is the opposite trend for reaction with benzyl azide. Introduction of electron-withdrawing substituents on the aryl azide led to a further acceleration in reaction with BCN, while reaction rate with DIBAC stayed more or less constant. The highest reaction rate acceleration was achieved with a pyridinium derivative, giving an almost 30-fold by faster reaction than benzyl azide and an absolute rate constant of almost 2 M^{-1} s^{-1} (Fig. 7).

3.4 Solvent Effects

Since Breslow first pointed out the effect of water on cycloaddition reactions [45], an extensive amount of kinetic measurements in the field have been determined, in particular focusing on the Diels–Alder reaction, a (4 + 2)-cycloaddition reaction. Studies of the comparative rates of 1,3-dipolar cycloaddition reactions in water and

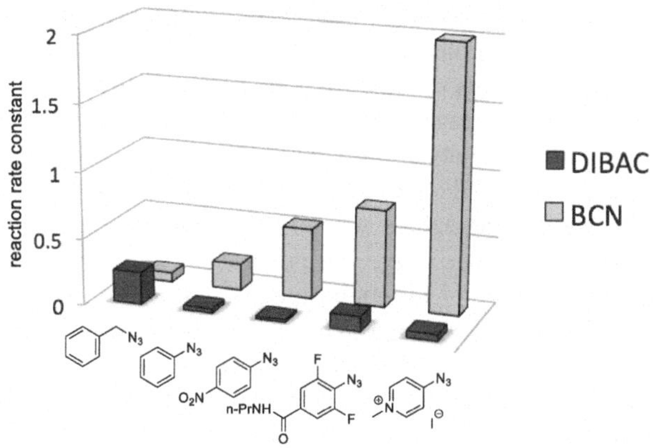

Fig. 7 Relationship between nature of the azide, structure of cyclooctyne and reaction rate constants. Reaction rate constants were determined in $CD_3CN:D_2O = 3:1$

organic solvents have not attracted the same level of attention. Some work in the field has focused on reactions of transient nitrile oxides [59] and nitrilimines [39] generated in aqueous environments and there are some reports on rate measurements on 1,3-dipolar cycloaddition reactions in water–organic mixtures. Recently, it was reported that with an increasing mole fraction of water, significant enhancement of 1,3-dipolar cycloaddition rates occurs [12]. It was suggested that a dominant hydrogen-bonding effect operates in water-induced rate enhancements of 1,3-dipolar cycloaddition reactions. The hydrogen-bonding effect involves secondary hydrogen bond bridging from the primary water-solvation shell of the transition state and the growth of structured water clusters, which was also supported by theoretical calculations.

The awareness that increasing levels of water translate into faster reaction rates has received surprisingly little attention in the field of SPAAC, despite the fact that a large number of applications involve conjugation processes on natural biomolecules (peptides, proteins, glycans, oligonucleotides) in aqueous systems. Nevertheless, classification and appreciation of a given cyclooctyne probe is typically only performed by determination of reaction rate constants in any of a range of solvents like methanol, acetonitrile or water (or deuterated versions thereof if performed in NMR, vide infra), or mixtures thereof. For example, Table 1 above displays the reaction rate constants of the known cyclooctynes across these solvents, which in fact makes it difficult to truly compare the usefulness of these probes in a quantifiable manner. Furthermore, it has been noted by several authors that SPAAC proceeds (significantly) faster in more aqueous solvent systems [9, 19, 24, 40, 60].

For these reasons, we embarked on a study to determine the reaction rate constants of the most commonly applied dibenzoannulated and aliphatic cyclooctynes, respectively DIBAC and BCN, as deduced from a SciFinder® structure search of these compounds. As becomes clear, since the year of its inception (2010), DIBAC has made its appearance in 260 unique publications (184 scientific publications and 80 patent applications) while BCN is reported 155 times (103 papers and 51 patent application). For comparison, DIBO, the third-most popular cyclooctyne gave 135 unique hits for the same time-frame (74 papers, 42 patent applications). Moreover, given the lower reaction rate constant of DIBO versus DIBAC, as well as their large structural similarity, DIBO was omitted from the studies below.

It was decided to evaluate both DIBAC and BCN with a range of representative azides and in three different solvent systems: MeOD, CD_3CN:D_2O (1:3), both for NMR measurements, and THF:H_2O (9:1), for quantification by IR. In order to ensure that both probes and azides would be soluble in the acetonitrile–water mixture, a hydroxyethylated derivative was prepared in several instances. The same strategy was applied to some of the more lipophilic aromatic azides, which led to full solubility except for the diisopropylated azidobenzene derivative (entry 5), where NMR was performed in a MeCN:H_2O (1:2) mixture.

Several interesting observations can be made from Table 2. First of all, the picture is confirmed that DIBAC is faster than BCN in reaction with aliphatic azides (entries 1–3). While apparent in all cases, the reaction rate difference is markedly larger in MeOD or CD_3CN:D_2O (factor ~10) than in THF:H_2O (factor ~3). Also striking is the rate constant of 1.9 M^{-1} s^{-1} for reaction of DIBAC with benzyl azide (entry 1), which is more than double than that for the other aliphatic azides (entries 2 + 3). We attribute this to the relatively low solubility of benzyl azide in an aqueous solvent system, resulting in faster reaction by means of a hydrophobic effect. The hydrophobic effect may also be accountable for another interesting observation, that the reactions of DIBAC with aliphatic azides are much faster in 75 % aqueous acetonitrile than in methanol or in 10 % aqueous THF.

As expected, the reactivity trend is reversed for azidobenzene (entry 4), where BCN is up to seven times faster than DIBAC (in 75 % aqueous CD_3CN), with a reaction a rate constant of 0.75 M^{-1} s^{-1}. As reported earlier [64], introduction of an *ortho*-isopropyl group on azidobenzene (entry 5) leads to significant rate acceleration for reaction with DIBAC, which is already very high in MeOD (2.3 M^{-1} s^{-1}) and too fast to measure in D_2O:CD_3CN by NMR. The number provided in the Table (~4 M^{-1} s^{-1}) for this solvent system is instead derived by multiplying the rate constant determined for BCN (1.5 M^{-1} s^{-1}) by the earlier determined relative reaction rate DIBAC:BCN = 2.5 [20]. A similar strategy estimates the reaction rate constant of BCN with the electron-poor difluorinated phenylazide (~6 M^{-1} s^{-1}, entry 6) from the value experimentally determined for DIBAC.

3.5 Tools to Quantify SPAAC Reaction Rates

The main determinant of the quality of any given cyclooctyne for SPAAC reaction is its reaction rate constant with azide (aliphatic or aromatic). Throughout the years, a large number of reaction rate constants have been experimentally determined for different cycloalkynes and azides, mainly by four analytical techniques: (1) NMR, (2) UV spectroscopy, (3) IR spectroscopy, and (4) fluorescence.

The most commonly applied method to determine a SPAAC reaction rate constant is by NMR [1]. To this end, a cyclooctyne and an azide are mixed in a deuterated solvent and formation of product is quantified by integration of diagnostic peaks of the formed triazole product. Given the fact that the triazole ring itself is fully substituted, other diagnostic protons in the product with a unique, non-

Table 2 Reaction rate constants for cycloadditions of aliphatic (entries 1–3) and aromatic (entries 4–6) azides with BCN and DIBAC in different solvent systems

Entry	Azide structure	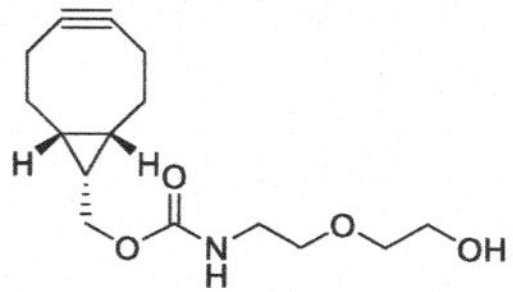			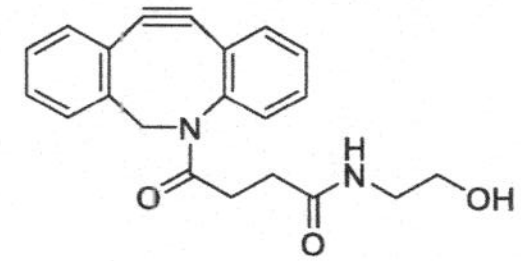		
		MeOD	D_2O:CD_3CN (3:1)	H_2O:THF (1:9)	MeOD	D_2O:CD_3CN (3:1)	H_2O:THF (1:9)
1	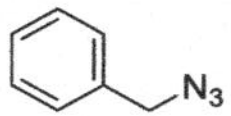	0.03	0.20	0.12	0.42	1.9	0.36
2	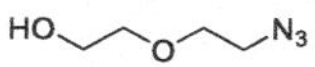	0.04	0.07	0.05[a]	0.33	0.90	0.21[a]
3	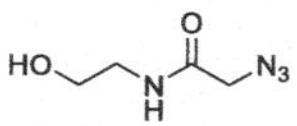	0.03	0.09	0.05[a]	0.29	0.82	0.20[a]
4	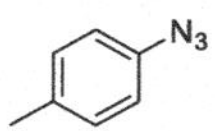	0.18	0.75	0.30	0.11	0.11	0.08
5	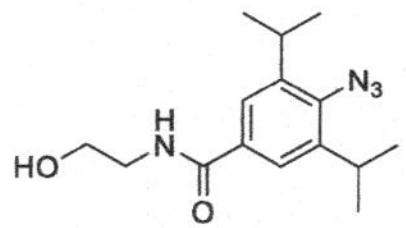	0.65	1.5[b]	0.30[a]	2.3	~4[b,c]	0.77[a]
6	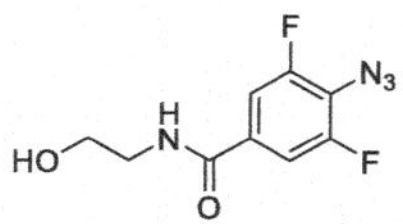	1.1	~6[c]	0.73[a]	0.18	0.95	0.11[a]

Rate constants in MeOD and D_2O:CD_3CN were determined by NMR, in H_2O:THF by IR

[a] Data taken from similar compounds, as reported earlier [20]

[b] In D_2O:CD_3CN (2:1)

[c] Estimated based on known ratio BCN:DIBAC [20]

overlapping shift in the NMR spectrum, must be identified (or specifically introduced in one of the substrates). Alternatively, one or more distinct proton peaks of starting material can be integrated. It must be noted that the formation of a mixture of regioisomeric triazoles, as is typical for all the cyclooctynes except the C_2-symmetric versions, is often a complicating factor. From the conversion plots thus obtained, the second-order rate plots can be calculated according to Eq. (1):

$$kt = \frac{1}{[B]_0 - [A]_0} \times \ln \frac{[A]_0([B]_0 - [P])}{([A]_0 - [P])[B]_0} \tag{1}$$

with k = second-order rate constant ($M^{-1}\ s^{-1}$), t = reaction time (s), $[A]_0$ = the initial concentration of substrate A (mmol/ml), $[B]_0$ = the initial concentration of substrate B (mmol/ml) and $[P]$ = the concentration of triazole product (mmol/ml). Either cyclooctyne or azide can be applied in excess, preferably between 1.2 and 2 (but not stoichiometric). By plotting t versus kt, calculation of the slope of the resulting (straight) line, gives the reaction rate constant k. Better plots are obtained by only including data points up to approximately 50 % conversion. Given the relatively high required concentration of components for fast NMR measurements (typically 10 mM or higher) and the rather time-consuming process before the first measurement can take place (mixing of reagents, insertion in magnet, shimming, scanning), it is clear that NMR has its limitations for very fast SPAAC (>0.5 $M^{-1}\ s^{-1}$), because more than 50 % of starting material may already be converted at the first measurement and few data points can be taken.

Two alternative techniques for real-time monitoring of (fast) SPAAC processes are UV and IR spectroscopy. For example, rate measurements of SPAAC can be conducted by reaction of a 10–100-fold excess of azide to a low concentration of acetylene (down to 1×10^{-4} M) in MeOH. Reaction in this case can be monitored by following the decay of the characteristic absorbance of the acetylene bond, as for example for dibenzoannulated cyclooctyne DIBO [43]. This method may give more accurate rate constants compared to the use of NMR, especially for fast reactions. Importantly, with this large excess of azide over cyclooctyne, the UV spectroscopic method is performed under pseudo-first-order conditions over a wide range of reagent concentrations, making the analysis of second-order kinetic curves more reliable. In case the UV absorption of the acetylene bond is less distinct, as for the aliphatic cyclooctynes, and no other specific UV chromophores can be identified in starting materials or reagents, an IR-based method may be a viable alternative. We recently reported [20] that the substrate-to-product conversion can be directly monitored by integration of the distinct stretch vibration of azide ($\sim$2100 cm^{-1}). Conveniently, deuterated solvents are not necessary and it was found that IR monitoring could even be executed in 10 % aqueous THF.

The most sensitive method for determination of reaction rate constants of cyclooctynes is by means of a reaction with a fluorogenic azide substrate [35]. By definition, a fluorogenic SPAAC process involves a reaction between a non-fluorescent alkyne and azide, allowing the ligation of two biomolecules to afford a highly fluorescent triazole product. Besides that, compounds that become fluorescent upon reaction with a chemical reporter and without the need of copper have many attractive features as bioorthogonal probes, such as eliminating the need for probe washout, reducing background labeling, and offering opportunities for monitoring biological processes in real time. The first fluorogenic click reaction based on readily synthesized azidocoumarin derivatives was reported by Sivakumar et al. [53]. Moreover, the photophysical fluorescent properties of coumarines can be strongly enhanced by substitution with an electron withdrawing group at the 3-position and/or an electron donating group at the 7-position. Other early variants of fluorogenic azides involve azidomethylated 1,8-naphthalimide dyes [48] and fluorogenic dyes based on a photoinduced electron transfer (PET) process with anthracene as a fluorophore [63]. It must be noted that the vast majority of fluorogenic probes feature an aromatic azide, which obviously will have a major impact on determination of reaction rate constant in absolute terms, as was for example found by us for a range of BCN derivatives [36]. Other more recent fluorogenic azides, as developed by Shieh et al. [51] and Herner et al. [26] and even a fluorogenic azide substrate for generation of a near-infrared (NIR) triazole dye [52], are depicted in Fig. 8.

Notwithstanding the elegancy and versatility of the azide-based fluorogenic probes, in particular to enable the visualization of biomolecules in living systems

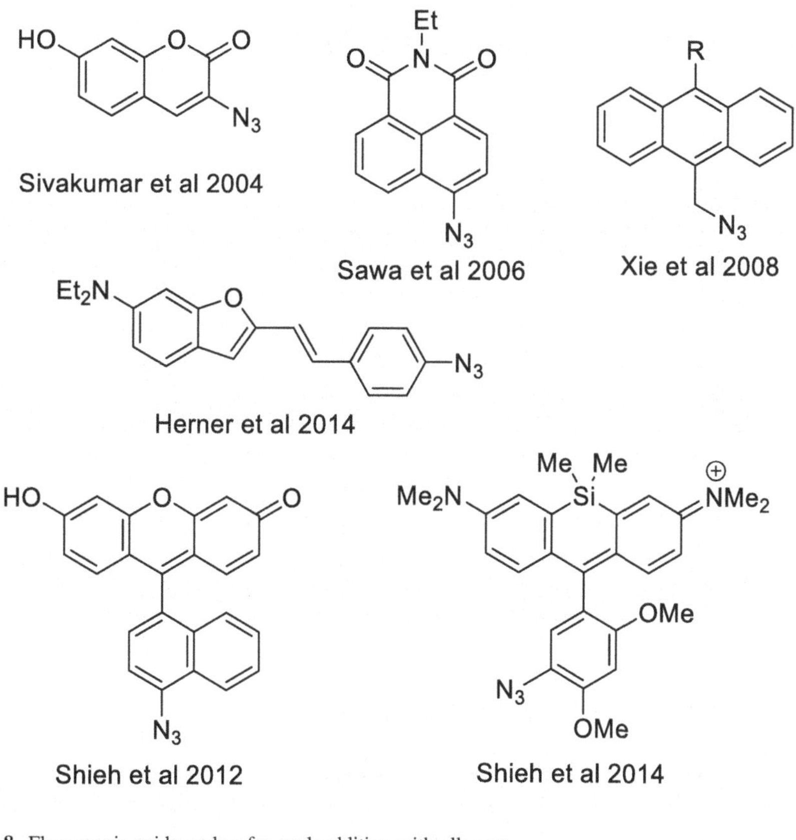

Fig. 8 Fluorogenic azide probes for cycloaddition with alkynes

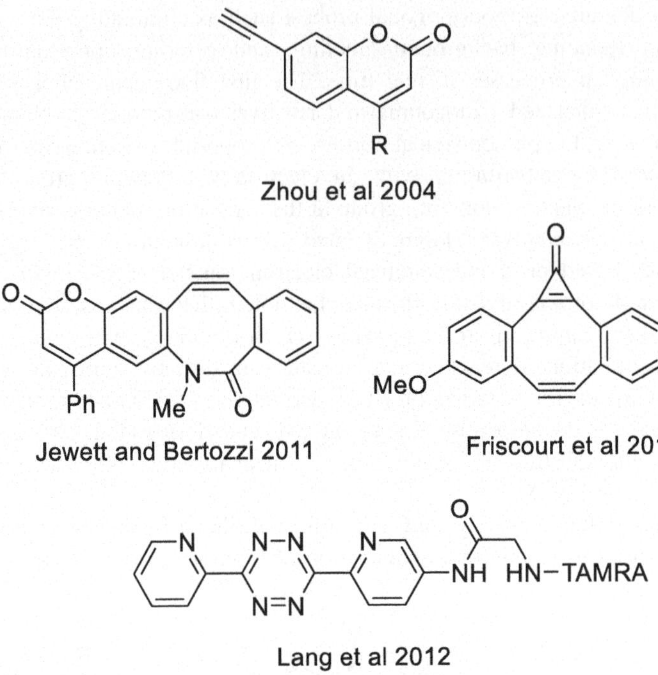

Fig. 9 Fluorogenic alkynes for cycloaddition with azides

with biooorthogonal chemistry, it is also clear that for fluorescence detection of an azide label in a biomolecular environment, a complementary fluorogenic alkyne structure is desired. The earliest example of the latter (Fig. 9) was provided in the form of acetylenic derivatives of coumarin by Zhou and Fahrni [65], highly analogous to the azido-coumarin derivatives described by Sivakumar et al. [53]. However, clearly the reaction of a terminal alkyne with an azide requires the undesirable presence of copper(I) to induce the formation of triazole. Two fluorogenic cyclooctyne versions have been developed in the past few years. The first probe was based on annulation of coumarin to BARAC and was developed by Jewett and Bertozzi [29]. Secondly, Friscourt et al. [23] reported that the cyclopropenone derivative of Sondheimer diyne unexpectedly forms a highly fluorescent triazole upon a copper-free reaction with azide. Most recently, Lang et al. [34] showed that a TAMRA-functionalized tetrazine derivative leads to fivefold to tenfold increase in fluorescence signal, similar to earlier reported for reaction with *trans*-cyclooctene and norbornene, upon reaction with BCN.

4 Concluding Remarks and Future Prospects

Strain-promoted azide–alkyne cycloaddition (SPAAC), since its inception in 2004, has firmly established itself as a powerful click chemistry tool. The commercial access of starting materials, its ease of operation, the nowadays practical reaction rates, even under high dilution conditions, and the stability of the resulting triazole product, have catapulted SPAAC at the forefront in many research areas in academia and industry. Originally developed for application in bioorthogonal chemistry, SPAAC has proven its value more and more in additional areas of science such as bioconjugation processes, hybrid and block polymers, high-performance and self-regenerative materials, metabolic engineering of biological systems and beyond. One key to the success of SPAAC is the azide component in the reaction, which is easily accessible, small and stable [17]. Nevertheless, recent years have elegantly demonstrated the power of cyclooctyne chemistry beyond cycloaddition with organic azides. Although much of these reactions were known for more than 50 years, mostly by work of Huisgen in the field of 1,3-dipolar cycloadditions, the past decade has witnessed a strong revival of cyclooctyne for development of fast and selective reaction with a range of other alkynophiles. For example, we were to apply [41] strain-promoted alkyne–nitrone cycloaddition (SPANC), a reaction also reported by McKay et al. in the same year [37], for the N-terminal labeling of proteins. Similarly, both our laboratory [27] and Sanders et al. [47] reported the reaction of cyclooctynes with nitrile oxides (SPANOC) leading to isoxazoles, which is a factor $\sim$10 faster than SPAAC and SPANC. Finally, strain-promoted cycloaddition with diazo-compounds is also known [47] as well as reaction of BCN with sydnone [42, 61]. Interestingly, in the field of (4 + 2) cycloadditions, cyclooctynes show reaction rate constants more than a factor 1000 higher than for 1,3-dipolar cycloadditions. It has been known for more than 30 years that aliphatic cyclooctynes undergo extremely fast (4 + 2) cycloadditions with electron-poor tetrazines [3]. The latter process can be tailored to specific reaction rates with electron-rich cyclooctynes like BCN in organic solvents [15] and was found to proceed at a surprisingly high reaction rates >1000 M^{-1} s^{-1} under aqueous conditions when applied for protein labeling [9, 34]. We most recently showed that (4 + 2) cycloaddition of BCN with 1,2-quinone (SPOCQ) proceeds with reaction rates intermediate of azide and tetrazine ($\sim$500 M^{-1} s^{-1}) and can be applied for rapid labeling of proteins with genetically encoded BCN [10] or formation of gel networks in the order of seconds [30].

Finally, it seems fair to state that SPAAC has now matured into more than just a spectacular click reaction or exciting research tool. As noted earlier, only in the past 5 years more than 100 patent applications have been filed on the use of SPAAC with DIBAC or BCN, with applications in e.g., nanoparticle functionalization, polymer functionalization, genetic encoding, (biodegradable) hydrogels, controlled drug release, oligonucleotide tagging, DNA libraries, peptide arrays, long-acting biopharmaceuticals, radioisotope labeling, and nuclear imaging. Among these, arguably the most prominent application of SPAAC is in the field of the selective and site-specific conjugation of proteins, such as PEGylation, radioisotope labeling, and controlled release of biopharmaceuticals. For example, in the field of targeted anti-cancer therapy alone, a range of pharmaceutical companies (MedImmune, Novartis, Agensys, Sutro, Innate Pharma, Synaffix) are building (part of) their technology around the controlled attachment of highly potent toxins (payloads) to monoclonal antibodies—to prepare antibody–drug conjugates or ADCs—by means of installation of azide into the protein (genetic encoding or enzymatically) followed by highly specific SPAAC. A remarkable position for the process of spontaneous cycloaddition of cyclooctyne and azide, discovered more than 60 years ago out of purely chemical curiosity.

References

1. Agard NJ, Prescher JA, Bertozzi CR (2004) A strain-promoted [3 + 2] azide-alkyne cycloaddition for covalent modification of biomolecules in living systems. J Am Chem Soc 126:15046–15047
2. Bailey AS and White JE (1966) Reaction of picryl azide with olefins. J Chem Soc B 819–822
3. Balcar J, Chrisam G, Huber FX, Sauer J (1983) Reactivity of nitrogen-heterocycles with cyclooctyne as dienophile. Tetrahedron Lett 24:1481–1484
4. Baskin JM, Prescher JA, Laughlin S et al (2007) Copper-free click chemistry for dynamic in vivo imaging. Proc Natl Acad Sci USA 104:16793–16797
5. De Almeida G, Sletten EM, Nakamura H, Palaniappan KK, Bertozzi CR (2012) Thiacycloalkynes for Copper-Free Click Chemistry. Angew Chem Int Ed 51:2443–2447
6. Bach RD (2009) Ring strain energy in the cyclooctyl system. The effect of strain energy on [3 + 2] cycloaddition reactions with azides. J Am Chem Soc 131:5233–5243
7. Blomquist AT, Burge RE Jr, Liu LH et al (1951) Many-membered carbon rings. IV. Synthesis of cyclononyne and cyclodecyne. J Am Chem Soc 73:5510–5511
8. Blomquist AT, Liu LH (1953) Many-membered carbon rings. VII. Cyclooctyne. J Am Chem Soc 75:2153–2154
9. Borrmann A, Milles S, Plass T (2012) Genetic encoding of a Bicyclo[6.1.0]nonyne-charged amino acid enables fast cellular protein imaging by metal-free ligation. ChemBioChem 13:2094–2099
10. Borrmann A, Fatunsin O, Dommerholt J et al (2015) Strain-promoted oxidation-controlled cyclooctyne-1,2-quinone cycloaddition (SPOCQ) for fast and activatable protein conjugation. Bioconj Chem 26:257–261
11. Brandsma L, Verkruijsse HD (1978) An improved synthesis of cyclooctyne. Synthesis. 290
12. Butler RN, Cunningham WJ, Coyne AG, Burke LA (2012) The influence of water on the rates of 1,3-dipolar cycloaddition reactions: trigger points for exponential rate increases in water-organic solvent mixtures. water-super versus water-normal dipolarophiles. J Am Chem Soc 126:11923–11929
13. Campbell-Verduijn LS, Mirfeizi L, Schoonen AK, Dierckx RA, Elsinga PH, Feringa BL (2011) Strain-promoted copper-free "Click" Chemistry for ^{18}F radiolabeling of Bombesin. Angew Chem Int Ed 123:11313–11316
14. Chadwick RC, Van Gyzen S, Liogier S, Adronov A (2014) Scalable synthesis of strained cyclooctyne derivatives. Synthesis 46:669–677
15. Chen W, Wang D, Dai C et al (2012) Clicking 1,2,4,5-tetrazine and cyclooctynes with tunable reaction rates. Chem Commun 48:1736–1738
16. Debets MF, van der Doelen CWJ, Rutjes FPJT et al (2010) Azide: a unique dipole for metal-free bioorthogonal ligations. ChemBioChem 11:1168–1184
17. Debets MF, van Berkel SS, Schoffelen S et al (2010) Aza-dibenzocyclooctynes for fast and efficient enzyme PEGylation via copper-free (3 + 2) cycloaddition. Chem Commun 46:97–99
18. Debets MF, Prins JS, Merckx D et al (2014) Synthesis of DIBAC analogues with excellent SPAAC rate constants. Org Biomol Chem 12:5031–5037
19. Dommerholt J, Schmidt S, Temming R et al (2010) Readily accessible bicyclononynes for bioorthogonal labeling and three-dimensional imaging of living cells. Angew Chem Int Ed 49:9422–9425
20. Dommerholt J, van Rooijen O, Borrmann A et al (2014) Highly accelerated inverse electron-demand cycloaddition of electron-deficient azides with aliphatic cyclooctynes. Nature Commun 5:5378

21. Domnin NA (1938) The triple bond in carbon rings and the probable structure of the simplest cyclic hydrocarbon of the composition C_nH_{2n-4}. II. Synthesis of cyclooctyne. J Gen Chem USSR 8:851–868
22. Ess DH, Jones GO, Houk KN (2008) Transition states of strain-promoted metal-free click chemistry: 1,3-dipolar cycloadditions of phenyl azide and cyclooctynes. Org Lett 10:1633–1636
23. Friscourt F, Ledin PA, Mbua NE et al (2012) Polar dibenzocyclooctynes for selective labeling of extracellular glycoconjugates of living cells. J Am Chem Soc 134:5381–5389
24. Gröst C, Berg T (2015) PYRROC: the first functionalized cycloalkyne that facilitates isomer-free generation of organic molecules by SPAAC. Org Biomol Chem 13:3866–3870
25. Heber D, Rösner P, Tochtermann W (2005) Cyclooctyne and 4-cyclooctyn-1-ol—versatile building blocks in organic synthesis. Eur J Org Chem 4231–4247
26. Herner A, Girona GA, Nikíc I (2014) New generation of bioorthogonally applicable fluorogenic dyes with visible excitations and large stokes shifts. Bioconj Chem 25:1370–1374
27. Jawalekar AM, Reubsaet E, Rutjes FPJT et al (2011) Synthesis of isoxazoles by hypervalent iodine-induced cycloaddition of nitrile oxides to alkynes. Chem Commun 47:3198–3200
28. Jewett JC, Sletten EM, Bertozzi CR (2010) Rapid cu-free click chemistry with readily synthesized biarylazacyclooctynones. J Am Chem Soc 132:3688–3690
29. Jewett JC, Bertozzi CR (2011) Synthesis of a fluorogenic cyclooctyne activated by cu-free click Chemistry. Org Lett 13:5937–5939
30. Jonker AM, Borrmann A, van Eck ERH et al (2015) A fast and activatable cross-linking strategy for hydrogel formation. Adv Mat 27:1235–1240
31. Krebs A, Kimling H (1971) [1 + 2] cycloadditions of isocyanides to alkynes. Angew Chem Int Ed 10:409–410
32. Krebs A, Wilke J (1983) Angle strained cycloalkynes. Top Curr Chem 109:189–233
33. Kuzmin A, Poloukhtine A, Wolfert MA, Popik VV (2010) Surface functionalization using catalyst-free azide-alkyne cycloaddition. Bioconj Chem 21:2076–2085
34. Lang K, Davis L, Wallace S et al (2012) Genetic encoding of bicyclononynes and trans-cyclooctenes for site- specific protein labeling in vitro and in live mammalian cells via rapid fluorogenic Diels–Alder reactions. J Am Chem Soc 134:10317–10320
35. Le Troumaguet C, Wang C, Wang Q (2010) Fluorogenic click reaction. Chem Soc Rev 39:1233–1239
36. Leunissen EHP, Meuleners MHL, Verkade JMM et al (2014) Copper-free click with polar BCN derivatives for modulation of cellular imaging. ChemBioChem 15:1445–1451
37. McKay CS, Moran J, Pezacki JP (2010) Nitrones as dipoles for rapid strain-promoted 1,3-dipolar cycloadditions with cyclooctynes. Chem Commun 46:931–933
38. Meier H (1972) Synthesis of medium-sized cycloalkynes. Synthesis 235–253
39. Molteni G, Orlandi M, Broggini G (2000) Nitrilimine cycloadditions in aqueous media. J Chem Soc Perkin Trans 1:3742–3745
40. Ning X, Guo J, Wolfert MA, Boons GJ (2008) Visualizing metabolically labeled glycoconjugates of living cells by copper-free and fast Huisgen cycloadditions. Angew Chem Int Ed 47:2253–2255
41. Ning X, Temming RP, Dommerholt J et al (2010) Protein modification by strain-promoted alkyne-nitrone cycloaddition. Angew Chem Int Ed 49:3065–3068
42. Plougastel L, Koniev O, Specklin S et al (2014) 4-Halogeno-sydnones for fast strain-promoted cycloaddition with bicyclo-[6.1.0]-nonyne. Chem Commun 50:9376–9378
43. Poloukthine AA, Mbua NE, Wolfert MA et al (2009) Selective labeling of living cells by a photo-triggered click reaction. J Am Chem Soc 131:15770–15776
44. Prelog V, Schenker K, Küng W (1953) Zur Kenntnis des Kohlen-stoffringes. 62. Mitteilung. Zur Kenntnis des Neuringes. Über die transanulare Oxydation der Cyclonene zu Cyclononandiolen-(1,5). Helv Chim Acta 36:471–482
45. Rideout DC, Breslow RJ (1980) Hydrophobic acceleration of Diels–Alder reactions. J Am Chem Soc 102:7816–7817
46. Rostovtsev VV, Green LG, Fokin VV et al (2002) A stepwise Huisgen cycloaddition process: copper(I)-catalyzed regioselective "ligation" of azides and terminal alkynes. Angew Chem Int Ed 41:2596–2599
47. Sanders BC, Friscourt F, Ledin PA et al (2011) Metal-free sequential [3 + 2]-dipolar cycloadditions using cyclooctynes and 1,3-dipoles of different reactivity. J Am Chem Soc 133:949–957
48. Sawa M, Hsu TL, Itoh T et al (2006) Glycoproteomic probes for fluorescent imaging of fucosylated glycans in vivo. Proc Nat Acad Sci USA 103:12371–12376

49. Saxon E, Bertozzi CR (2000) Cell surface engineering by a modified Staudinger reaction. Science 287:2007–2010
50. Scheiner P, Schomaker JH, Deming S et al (1965) The addition of aryl azides to norbornene. A kinetic investigation. J Am Chem Soc 87:306–311
51. Shieh P, Hangauer MJ, Bertozzi CR (2012) Fluorogenic azidofluoresceins for biological imaging. J Am Chem Soc 134:17428–17431
52. Shieh P, Sloan Siegrist M, Cullen AJ et al (2014) Imaging bacterial peptidoglycan with near-infrared fluorogenic azide probes. Proc Nat Acad Sci 111:5456–5461
53. Sivakumar K, Xie F, Cash BM et al (2004) A fluorogenic 1,3-dipolar cycloaddition reaction of 3-Azidocoumarins and acetylenes. Org Lett 6:4603–4606
54. Sletten EM, Bertozzi CR (2008) A hydrophilic azacyclooctyne for Cu-free click Chemistry. Org Lett 10:3097–3099
55. Sletten EM, Nakamura H, Jewett JC et al (2010) Difluorobenzocyclooctyne: synthesis, reactivity, and stabilization by β-cyclodextrin. J Am Chem Soc 132:11799–11805
56. Stöckmann H, Neves AA, Stairs S et al (2011) Development and evaluation of new cyclooctynes for cell surface glycan imaging in cancer cells. Chem Sci 2:932–936
57. Tornøe CW, Christensen C, Meldal M (2002) Peptidotriazoles on solid phase: [1,2,3]-triazoles by regiospecific Copper(I)-catalyzed 1,3-dipolar cycloadditions of terminal alkynes to azides. J Org Chem 67:3057–3064
58. Turner RB, Jarrett AD, Goebel P et al (1973) Heats of hydrogenation. IX. Cyclic acetylenes and some miscellaneous olefins. J Am Chem Soc 95:790–792
59. Van Mersbergen D, Wijnen JW, Engberts JBFN (1998) 1,3-dipolar cycloadditions of benzonitrile oxide with various dipolarophiles in aqueous solutions. A kinetic study. J Org Chem 63:8801–8805
60. Varga BR, Kállay M, Hegyi K et al (2012) A non-fluorinated monobenzocyclooctyne for rapid Copper-free click reactions. Chem Eur J 18:822–828
61. Wallace S, Chin JW (2014) Strain-promoted sydnone bicyclo-[6.1.0]-nonyne cycloaddition. Chem Sci 5:1742–1744
62. Wittig G, Krebs A (1961) On the existence of small-membered cycloalkynes, I. Chem Ber 94:3260–3275
63. Xie F, Sivakumar K, Zeng Q et al (2008) A fluorogenic Cu(I) catalyzed azide-alkyne cycloaddition reaction of azidoanthracene derivatives. Tetrahedron 64:2906–2914
64. Yoshida S, Shiraishi A, Kanno K et al (2011) Enhanced clickability of doubly sterically-hindered aryl azides. Sci Rep 82:1–4
65. Zhou Z, Fahrni CJ (2004) A fluorogenic probe for the Copper(I)-catalyzed azide-alkyne ligation reaction: modulation of the fluorescence emission via $^{3}(n,\pi^{*})$–$^{1}(\pi,\pi^{*})$ inversion. J Am Chem Soc 126:8862–8863
66. Zimmerman ES, Heibeck TH, Gill A et al (2014) Production of site-specific antibody–drug conjugates using optimized non-natural amino acids in a cell-free expression system. Bioconj Chem 25:351–361

Top Curr Chem (Z) (2016) 374:1
DOI 10.1007/s41061-015-0002-2

REVIEW

Photo-Triggered Click Chemistry for Biological Applications

András Herner[1] · Qing Lin[1]

Received: 28 October 2015 / Accepted: 11 November 2015 / Published online: 11 December 2015

Abstract In the last decade and a half, numerous bioorthogonal reactions have been developed with a goal to study biological processes in their native environment, i.e., in living cells and animals. Among them, the photo-triggered reactions offer several unique advantages including operational simplicity with the use of light rather than toxic metal catalysts and ligands, and exceptional spatiotemporal control through the application of an appropriate light source with pre-selected wavelength, light intensity and exposure time. While the photoinduced reactions have been studied extensively in materials research, e.g., on macromolecular surface, the adaptation of these reactions for chemical biology applications is still in its infancy. In this chapter, we review the recent efforts in the discovery and optimization the photo-triggered bioorthogonal reactions, with a focus on those that have shown broad utility in biological systems. We discuss in each cases the chemical and mechanistic background, the kinetics of the reactions and the biological applicability together with the limiting factors.

Keywords Bioorthogonal reaction · Photo-triggered reaction · Photoclick · Tetrazole · Nitrile imine · Azirine · Cyclopropenone · *o*-Naphthoquinone methide · *o*-Quinodimethanes · Hetero Diels–Alder reaction

Abbreviations

2PE	Two-photon excitation
CuAAC	Cu-catalyzed azide–alkyne cycloaddition
DA	Diels–Alder
ε	Absorption coefficient

✉ Qing Lin
qinglin@buffalo.edu

[1] Department of Chemistry, State University of New York at Buffalo, Buffalo, NY 14260, USA

Φ	Quantum yield
HOMO	Highest occupied molecular orbital
λ	Wavelength
LUMO	Lowest unoccupied molecular orbital
PB	Phosphate buffer
PBS	Phosphate buffered saline
PG	Protecting group
SPAAC	Strain-promoted azide–alkyne cycloaddition
UAA	Unnatural amino acid
UV	Ultraviolet

1 Introduction

To study biological pathways in living system, it is often necessary to fluorescently label the biomolecules involved in the pathway so that their movement and function can be visualized directly inside a living cell. To this end, click chemistry [1] and many other bioorthogonal reactions have been developed in the last decade and a half, and they all share the following features: (1) the reactants do not interfere with the native biochemical processes inside cells; (2) the reaction rate is comparable to that of a biological process, and (3) the reactants are stable and nontoxic in living cells and animals [2, 3]. Major efforts have been devoted to achieve a faster reaction rate, to improve the physicochemical properties of the reactants such as size, solubility and membrane permeability, and to demonstrate the utilities of these

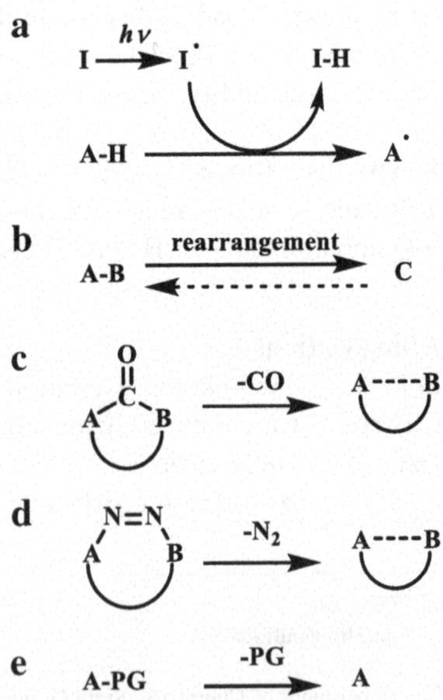

Scheme 1 Photo-triggered generation of reactive intermediates

reactions in biological systems. Amongst these reactions, the photo-triggered reactions represent a special class as they generally exhibit a finer control of the reaction initiation and duration, together with greater resolution in space and time. These controls are achieved through the use of an appropriate light source with pre-selected wavelength, light intensity and exposure time. Because the photo-triggered reactions proceed without the need of metal catalysts and ligands, they are generally less toxic to cells than the metal-catalyzed reactions.

The role of light in the photo-triggered cycloaddition reactions is to generate the reactive species, which then react with their suitable partners present in the biological systems. There are five ways to generate the reactive species through photo irradiations (Scheme 1), including: (a) the formation of radicals with a photo initiator (e.g., from thiols [4, 5]; (b) the rearrangement (e.g., *ortho*-quinodimethanes from *ortho*-methyl phenyl ketones/aldehydes [6], *ortho*-quinone methides from ortho-hydroxibenzyl alcohols [7], azirines [8]; (c) the expulsion of CO (e.g., cyclopropenones [9]); (d) the expulsion of N_2 (e.g., tetrazoles); and (e) the removal of photo-cleavable groups (e.g., nitrobenzyls, nitroaryls, coumarins) [10] (Scheme 1). In our review, we will focus on the reactive species that participate in the cycloaddition reactions after their photo-generation.

Depending on their stability, the photo-generated reactive species serving as the cycloaddition partners can be classified into two categories: (a) the reactive species are intrinsically unstable and as a result, they cannot be isolated as discrete compounds using traditional techniques and used as such in the cycloaddition reaction in biological environment (Scheme 2a); and (b) the reactive species are stable but their reactivity are masked in a protected form, which can be unmasked through a burst of light (Scheme 2b). In both cases, the triggering of the downstream cycloaddition reaction requires the photo activation of the precursors, yielding a photo control over the entire process. However, significant differences exist between these two processes: in the former, the photo-generation of the reactive species is irreversible (usually with expulsion of the CO or N_2 gas), and the half-life of the unstable reactive species can be tuned through structural modifications; in the latter, the stable intermediate can be reverted back to the unreactive "masked" form through chemical transformations. Furthermore, the long half-life of reactive species in the latter may reduce the spatial resolution as they

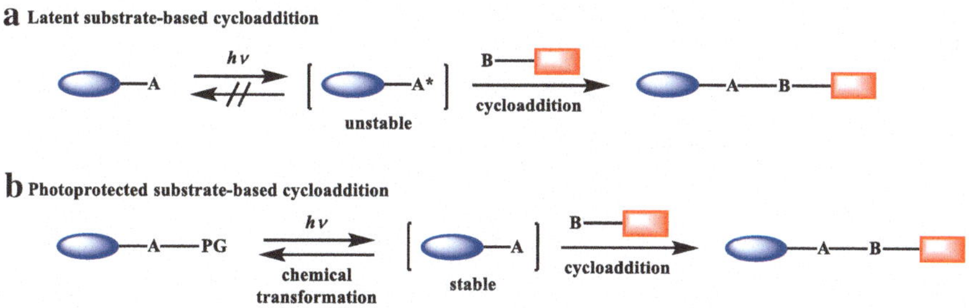

Scheme 2 **a** Strategy based on photoactivation of the precursor to generate a highly reactive and unstable species. **b** Strategy based the removal of a photo-protecting group

may diffuse away from the desired subcellular location before encountering its cycloaddition reaction partner in cellular applications.

From the cycloaddition reaction standpoint, the photo-generated reactive species suitable for both 1,3-dipolar cycloaddition reactions and Diels–Alder reactions have been studied in the literature. In this review, we will focus our discussion on four cycloaddition reactions: (1) 1,3-dipolar cycloaddition between alkenes and photo-generated nitrile imines; (2) 1,3-dipolar cycloaddition between alkenes and photo-generated nitrile ylides; (3) photoinduced hetero Diels–Alder reactions; and (4) photoinduced strain promoted azide–alkyne cycloaddition (Scheme 3). It is

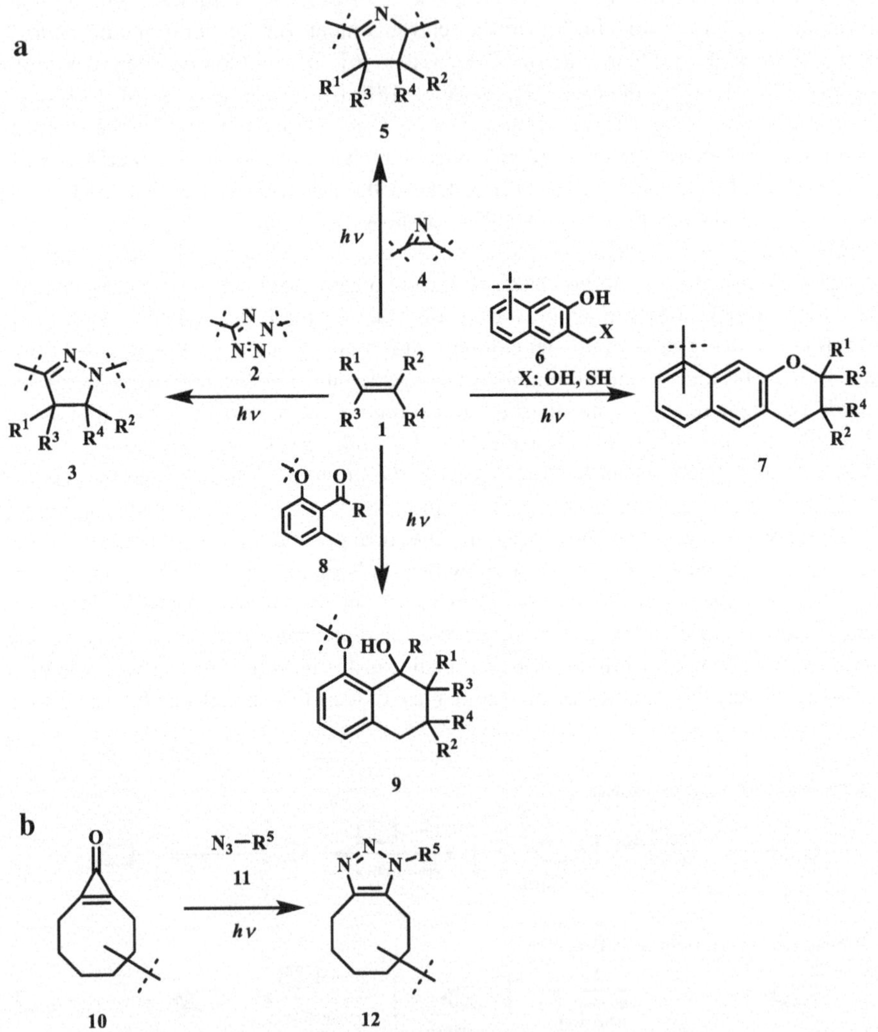

Scheme 3 Summary of photo-triggered cycloaddition reactions useful in biological applications. **a** Reactions between alkenes and photochemically generated 1,3-dipoles or 1,3-dienes. **b** SPAAC between photoprotected cyclooctynes and azides

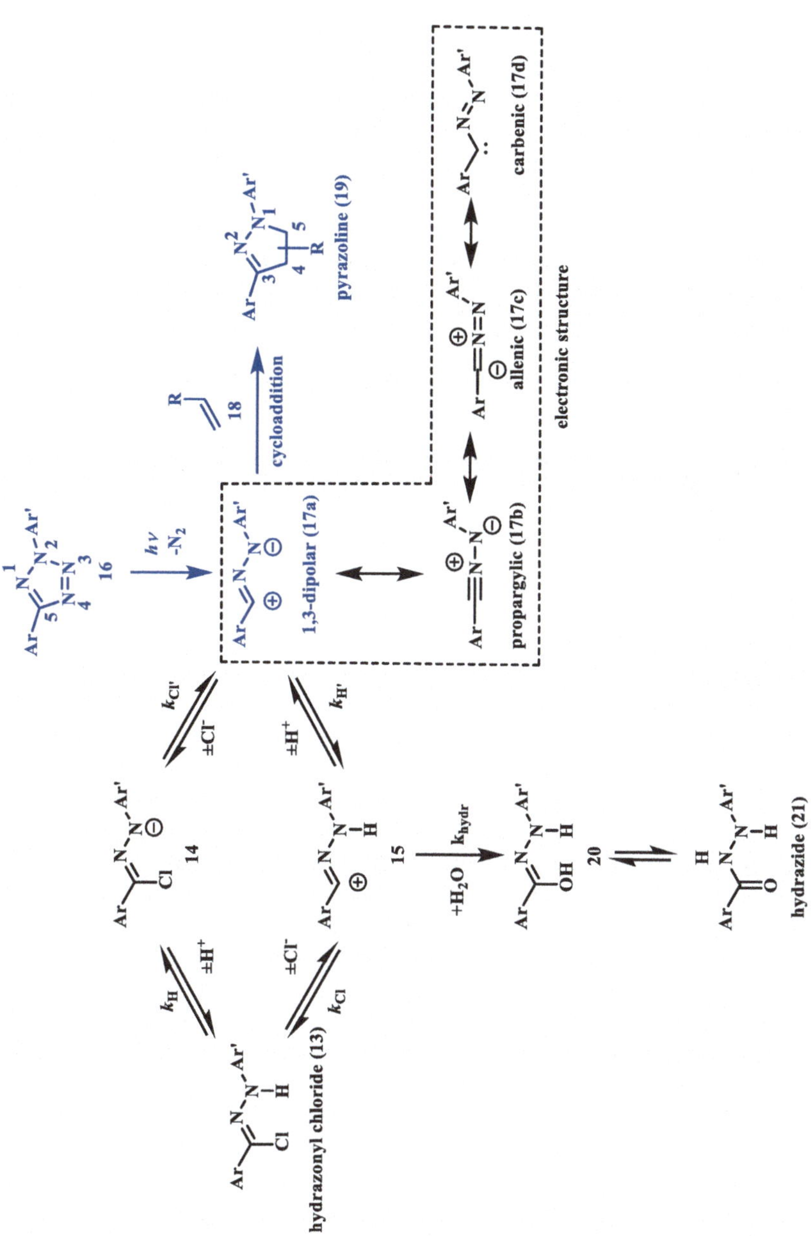

Scheme 4 Plausible mechanistic pathways for the generation of nitrile imine and its reactions in the chloride-containing aqueous medium. The photoinduced tetrazole ring rupture followed by cycloaddition with alkenes are colored in *blue*

noteworthy that the majority of photo-generated reactive species serve as either 1,3-dipoles or 1,3-dienes and participate the cycloaddition reactions with appropriately functionalized alkene, presumably due to the lower activation barriers in the concerted processes [11]. Specifically, the following aspects of the reactions will be discussed: (1) the chemical structure, synthesis and substituent effects on reacting partners; (2) the byproducts and possible side products; (3) photophysical parameters including photo-activation wavelength, light intensity, exposure time and quantum yield; (4) the kinetics of the cycloaddition reaction; and (5) the applications in biological systems and potential limiting factors.

2 Photoinduced 1,3-Dipolar Cycloaddition Reaction Between Tetrazoles and Alkenes

2.1 Theoretical Background and Reaction Mechanism

In the 1960s Huisgen discovered that the photo- or thermo-lysis of 2,5-diphenyl tetrazole (**16**) and treatment of hydrazonyl chlorides (**13**) under basic conditions led to the same pyrazoline cycloadduct (**19**) when methyl crotonate was used as the dipolarophile [12, 13]. The product was a 3:1 isomeric mixture of pyrazolines. After photoirradiation and N_2 release, the reactive nitrile imine (**17a**) is generated from tetrazole (**16**) irreversibly. In contrast to other bioorthogonal reactions whereby 1,3-dipoles are generally stable in aqueous medium (e.g., azide and nitrone), nitrile imine reacts with water and therefore needs to be formed in situ. A concerted reaction mechanism was proposed; first the diaryltetrazole (**16**) undergoes facile cycloreversion and generates in situ a nitrile imine dipole (**17a**), which reacts spontaneously with dipolarophiles (e.g., a suitable alkene, **18**) to afford the pyrazoline (**19**) product. The presence of the short-lived nitrile imine was verified by the fragmentation study of the ^{15}N enriched tetrazoles at low temperature [14]. A bent nitrile imine geometry was observed directly after photo-irradiating a Zn coordination crystal, and the bent geometry suggested the 1,3-dipolar electronic form (**17a**) (in dashed box in Scheme 4) based on the water-quenching study [15]. For diaryltetrazoles, the photoinduced ring rupture was very efficient in the middle UV region (<290 nm) with a quantum yield of 0.5–0.9 and the substituent effect was not significant [16, 17]. The experimental observation of the major product of the 5-pyrazoline was also supported by theoretical calculations in which the transition state was examined [18]. The cycloaddition is faster in water thanks to the hydrophobic effect [19]. Despite these favorable features, tetrazoles were used mostly in the synthesis of heterocyclic compounds [20, 21] and in polymer and material sciences [22] for a long time.

In addition to the photo-generation, Carell and coworkers showed that the hydrazonyl chloride (**13**) could also serve as a precursor to the reactive nitrile imine (**17a**) in neutral aqueous buffer [23]. Intrigued by these results, Liu and coworkers conducted a mechanistic study of the hydrazonyl chloride-mediated cycloaddition by examining the effects of pH and chloride concentration on the rates of nitrile imine formation and subsequent cycloaddition reactions [24]. They concluded that the cycloaddition reaction is most favorable at basic pH and in the absence of the chloride ion, and that the cycloaddition rate approaches $3.4 \times 10^4\ M^{-1}\ s^{-1}$ after factoring in the various equilibria, which makes this cycloaddition one of the fastest click reactions. A plausible mechanism unifying the two nitrile imine generation pathways is shown in Scheme 4, in which water or HCl addition products serve as reservoirs for the unstable nitrile imine prior to the cycloaddition reaction.

2.2 Synthesis of Tetrazoles

Two common synthetic routes have been used for the synthesis of tetrazoles (Scheme 5). First, the phenyl sulfonylhydrazones (**23**) react with the freshly prepared diazonium salts (**24**) in pyridine under cooled conditions to produce the desired tetrazole products (**25**) with moderate to good yields [25]. The hydrazones (**23**) can be synthesized easily from the aliphatic or aromatic aldehydes (**21**) and the diazonium salts from the corresponding amines with $NaNO_2$ under acidic conditions. The second method has been used to prepare the thiophene conjugated tetrazoles (**29**) which exhibit longer photoactivation wavelength [26–28].

2.3 Effects of Substrates on Photoactivation Wavelength and Spectral Properties of Pyrazoline

In 2007 we started to study tetrazoles for both synthetic and biological applications. In our first report [29], different solvent systems ranging from nonpolar benzene to $EtOH/H_2O$ (7:3) mixture were investigated along with the study of substituent effect on the phenyl rings (**30**) (Scheme 6a). The photoinduced cycloaddition reaction was found to be very robust with >90 % yields for most substrates along with high solvent tolerance and exclusive regioselectivity with the electron-withdrawing group residing at the C^5-position. In addition, it was found that a simple hand-held

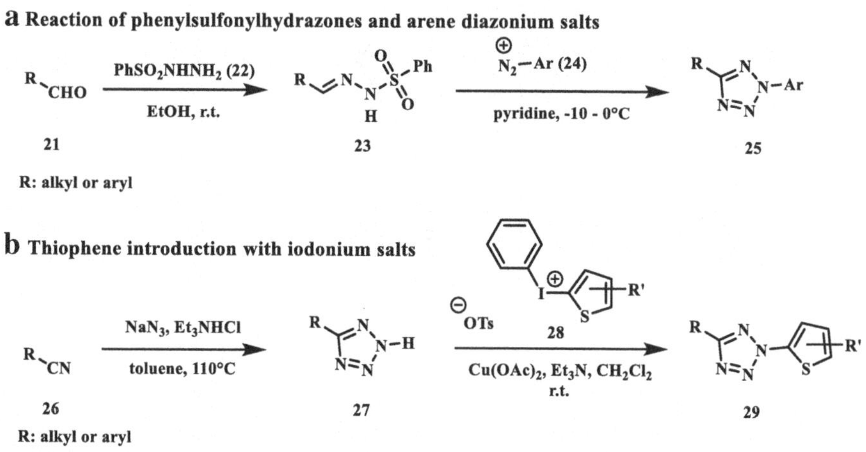

Scheme 5 Synthesis of *N*-aryltetrazoles

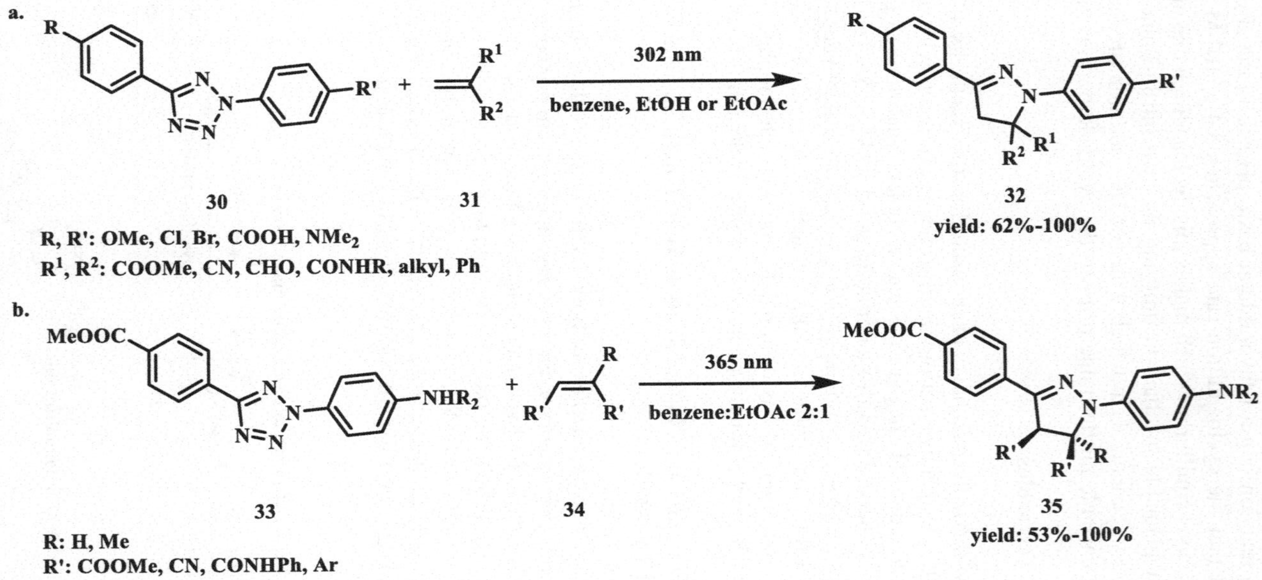

Scheme 6 Reaction of biaryltetrazoles with various dipolarophiles in the photoinduced cycloaddition reaction: **a** first study, **b** optimizing the photoactivation wavelength

benchtop UV lamp (UVP, 302 nm, 0.16 AMPS, typically used in the lab for thin layer chromatograph monitoring) was sufficient for driving the reaction to completion, making the reactions very accessible to most chemistry labs.

Since 302-nm UV light may pose considerable phototoxicity to cells [30], for wider use of the tetrazole-based cycloaddition reaction in biological systems it is imperative to shift the photoactivation wavelength to the long-wavelength region. To this end, a series of substituted diaryltetrazoles (**33**) were synthesized (Scheme 6b) [31], and their absorption maxima and absorption coefficients at 365 nm (wavelength of the long-wavelength hand-held UV lamp) were determined in MeOH/H_2O (1:1). It was found that the electron-donating NH_2 and NMe_2 groups yielded the largest shift in absorption maxima (to 310 and 336 nm, respectively) along with increased absorption coefficients at 365 nm (0.35×10^4 and 1.87×10^4 M^{-1} cm^{-1}, respectively).

Additional diaryltetrazole derivatives with potential long wavelength photoactivatability were designed and synthesized based on the 'scaffold hopping' strategy (Scheme 7, compounds **36–40**) [32]. Extension of the aromatic conjugation system provided a bathochromic shift of the photoactivation wavelength. These tetrazoles underwent efficient ring rupture upon 365 nm photoirradiation and generated the brightly fluorescent pyrazoline products after the cycloaddition reaction, which could be useful for visualization of the alkene-tagged proteins in living cells. Separately, appending oligothiophenes to the tetrazole core provided further shift in absorption maxima close to 405 nm, a wavelength for a commonly used laser source on microscopes (Scheme 7, compounds **41–45**) [28]. The quantum yield for 405 nm laser-induced ring rupture was determined to be 0.16, significantly higher than the reported 365-nm photoactivatable tetrazoles ($\Phi = 0.006$–0.04) [31]. A water-soluble derivative of **43**, compound **55**, exhibited second order rate constant, k_2, of 1299 ± 110 M^{-1} s^{-1} in the cycloaddition reaction with mono-methyl fumarate amide in MeCN/PBS (1:1). In addition, compound **55** was shown to selectively image the microtubules in CHO cells in a spatiotemporally controlled manner when the cells were pre-treated a fumarate modified docetaxel **56** (Fig. 1). To fine-tune the emission wavelength of the resulting pyrazoline adducts, additional dithiophene-containing tetrazole analogs were prepared (Scheme 7, compounds **46–54**) [33]. These compounds showed fast kinetics in the reaction with dimethyl fumarate (k_2 up to 3960 M^{-1} s^{-1} in 1:1 PBS/MeCN) upon 405-nm photoirradiation, and most importantly, provided the red fluorescent pyrazoline adducts (emission maxima in the range of 575–644 nm in 1:1 PBS/MeCN) in excellent yields.

To reduce light scattering and improve three dimensional localization of excitation, the two-photon excitation (2PE) of tetrazoles was demonstrated with the naphthalene-based derivatives [34]. By taking advantage of the known 2PE of naphthalene [35], various naphthalene-based tetrazole derivatives were synthesized and tested (**57–59**) (Fig. 2). A femtosecond 700 nm near infrared (NIR) pulsed laser source was used for initiating the photoclick chemistry in this study. The two-photon absorption cross section (δ_{aT}) of tetrazole **59** was determined to be 12 GM ($=10^{-50}$ cm^4 s/photon) and the cycloaddition cross section was 3.8 GM, comparable to the uncaging efficiency of the commonly used two-photon protecting group, 6-bromo-7-hydroxycoumarin-4-ylmethyl acetate ($\delta_{ur} = 0.95$ GM), under similar conditions [36]. Under the 2PE conditions, tetrazole **59** was used successfully in the labeling of the acrylamide-tagged GFP in PBS in vitro as well as the imaging of the microtubules in a spatiotemporally controlled manner in live CHO cells treated with the fumarate-modified docetaxel **56** (Fig. 2).

2.4 Site-Specific Labeling of Proteins Via the Photoinduced Tetrazole–Alkene Cycloaddition

One of the important applications of bioorthogonal reactions is labeling of proteins carrying amino acid side chains with unique chemical reactivity in their native environment [37]. This is usually accomplished in two steps: (1) a bioorthogonal reporter is introduced into the protein of interest; and (2) a biophysical probe with the cognate chemical reactivity reacts with the pre-tagged protein selectively. For the photoclick chemistry, the introduction of an alkene or a tetrazole can be achieved using either in vitro modification of the native protein side chains (e.g., Cys and Lys) or unnatural amino acids (UAAs) carrying the suitable functionality in vivo (Scheme 8) [38]. In general, the genetic approach in encoding of UAAs has overcome many disadvantages associated with the native residue-based chemistry, e.g., low selectivity, and as a result, has gained increasing popularity in protein science whenever specific protein modifications are needed, both in vitro and in vivo.

In the very first application of the tetrazole-based cycloaddition chemistry to proteins [39], a carboxylic acid functionalized tetrazole was coupled to a tripeptide (RGG) and the kinetics of the cycloaddition reaction between the tetrazole-modified peptide and acrylamide was investigated under the 302-nm photoirradiation condition. The photolysis of the tetrazole-modified peptide to its corresponding nitrile imine was extremely rapid with a first-order rate constant to be 0.14 s^{-1}; the subsequent cycloaddition with acrylamide proceeded with a second-order rate constant, k_2, of 11.0 M^{-1} s^{-1}. In the next step, the surface Lys residues of lysozyme were modified with a water-soluble tetrazole succinimide (**60**), and the resulting tetrazole-modified lysozyme was irradiated with 302 nm light for 2 min in the presence of acrylamide (Scheme 8a). The LC–MS analysis indicated that the conversion of the tetrazole-modified lysozyme to the pyrazoline adduct was very specific with an estimated yield around 90 %. We also prepared a tetrazole-containing enhanced green fluorescent protein, EGFP-Tet, using the intein-based chemical ligation strategy (Scheme 8a). The photo-triggered cycloaddition reaction between EGFP-Tet and *N*-hexadecylmethacrylamide proceeded in bacterial lysate with ~52 % conversion based on LC–MS analysis. The application of the tetrazole-modified EGFP was also used to probe the effect of lipidation on protein localization in live cells without the use of lipidation enzymes [40].

The tetrazole-based cycloaddition chemistry was subsequently applied to protein labeling in bacteria cells via a genetically encoded *O*-allyl-tyrosine (***O*-allyl-Tyr**, Scheme 8c) [41]. In this study, the reactions between tetrazole derivatives and allyl phenyl ether were first investigated. The formation of nitrile imine was found to be fast (<2 min) but the cycloaddition steps was slow ($k_2 = 0.00202$ M^{-1} s^{-1} with allyl phenyl ether in 1:1 PBS/MeCN; 75-fold slower than acrylamide). In-gel

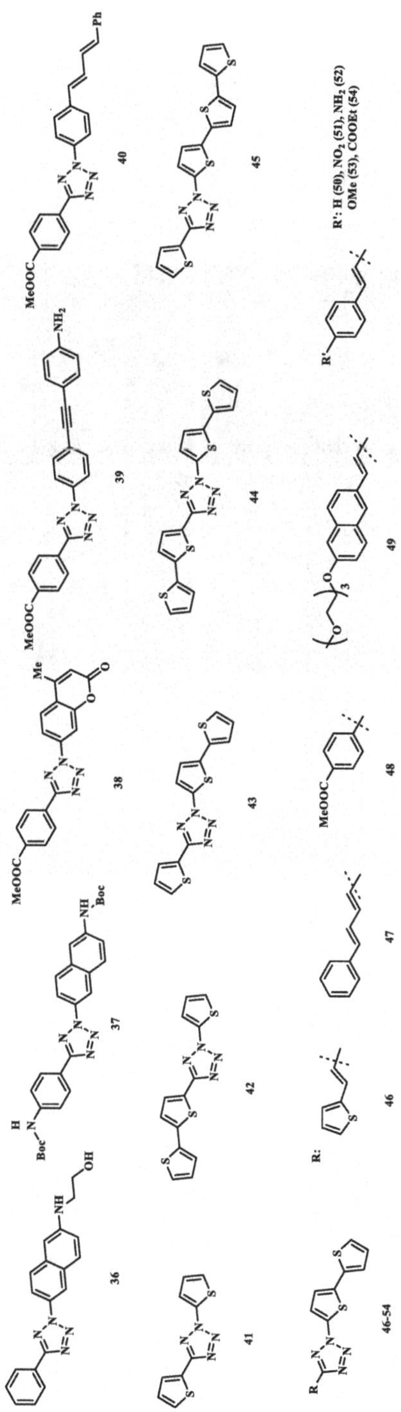

Scheme 7 Tetrazoles with the extended π-systems

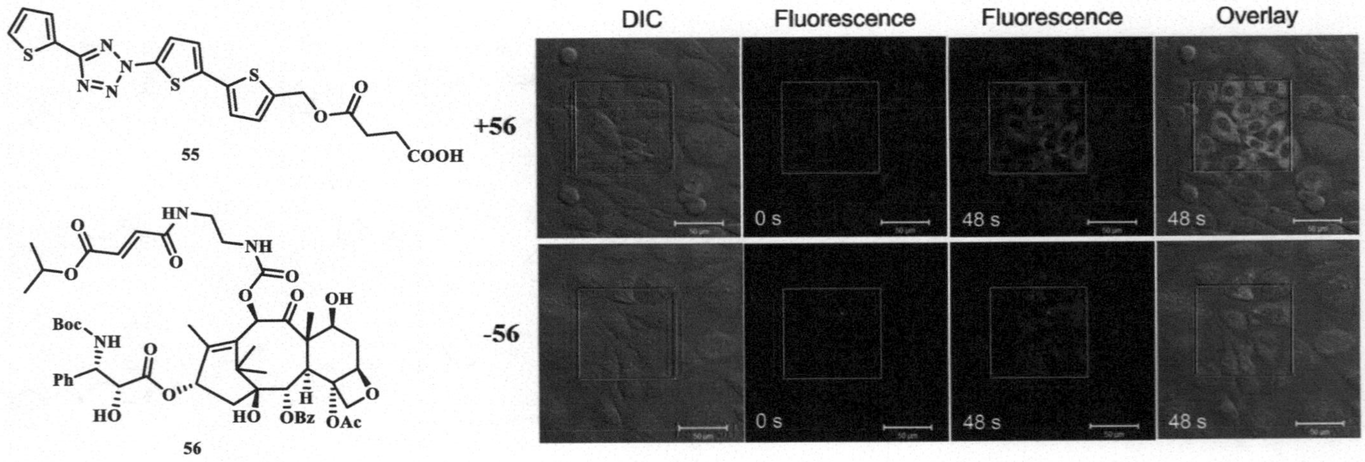

Fig. 1 Fluorescent imaging of microtubules in CHO cells using 405 nm photoactivatable tetrazole **55** in the presence or in the absence of a fumarate-modified docetaxel **56**. The *rectangle area* were illuminated with the 405-nm laser for the indicated time

fluorescence-based screen identified 4-(methoxycarbonyl)phenyl substituted tetrazole in the C^5-position from a panel of tetrazole compounds as a suitable reagent for selective labeling of *O*-allyl-tyrosine-encoded Z-domain protein, which was then used to image the same protein in *E. coli* cells. Since the reaction rate is inversely related to the energy gap between $HOMO_{dipole}$ and $LUMO_{dipolarophile}$ in the nitrile imine–alkene cycloaddition [18], it was envisioned that raising the HOMO energy of the nitrile imine should lead to reaction acceleration. Thus, a series of substituted diaryltetrazoles were synthesized and their reaction rates toward 4-penten-1-ol were measured [42]. It was found that the electron-donating substituents on the N^2-phenyl ring such as $–NH_2$ and –OMe groups significantly raise the HOMO energies of the photo-generated nitrile imine dipoles, with second-order rate constants reaching as high as 0.79 M^{-1} s^{-1} in 1:1 PBS/MeCN. Furthermore, a methoxy-substituted diphenyltetrazole was shown to label the *O*-allyl-tyrosine-encoded Z-domain proteins in *E. coli* in <1 min.

To overcome slow reaction kinetics observed with the genetically encoded system, two complementary efforts were untaken. One approach was to incorporate the tetrazole moiety into the amino acid side chain and evolve orthogonal aminoacyl-tRNA synthetase/tRNA pairs to charge the resulting tetrazole amino acids into proteins site-specifically. To this end, we synthesized a series of tetrazole modified unnatural amino acids (Scheme 8b), and tested their reactivity in the photo-triggered cycloaddition reaction with methyl methacrylate in PBS/MeCN (1:1) [43]. While four out of six tetrazole amino acids gave excellent reaction yields, only *p*-tetrazole-phenylalanine (***p*-Tpa**) was incorporated into myoglobin site-specifically using an evolved synthetase via the amber codon suppression [44]. A drawback of this genetically encodable tetrazole amino acid is that a 254-nm UV light is needed for triggering the reaction because of the shortened π-conjugation system. In addition, the cycloaddition reaction with dimethyl fumarate proceeded rather sluggishly, with a measured k_2 value of 0.082 ± 0.011 M^{-1} s^{-1} in 1:1 PBS/MeCN.

The second approach focused on the genetic encoding of the reactive alkene-containing amino acids, which are in general smaller than the tetrazole amino acids and thus easier to derive the specific orthogonal aminoacyl-tRNA synthetase/tRNA pairs (Scheme 8c). To increase alkene reactivity in the photoclick chemistry without unwanted side reactions such as Michael addition, ring strain has been successfully exploited. In 2010, we showed that norbornene served as an excellent substrate in the photo-triggered cycloaddition reaction with the macrocyclic tetrazoles [45]. Later, Carell and co-workers showed that a norbornene-modified Lys (**68**) can be genetically encoded into His_6-tagged human polymerase κ, and that efficient labeling of polymerase κ can be accomplished with the nitrile imines generated from either the hydrazonyl chloride precursor or the corresponding tetrazole [23]. The utility of the norbornene modified amino acids was also demonstrated in the tetrazine ligation in eukaryotic cells by other research groups [46–48]. Smaller strained alkenes such as cyclopropene have also been developed for the photoclick chemistry. Cyclopropene has a ring strain of 54.1 kcal mol^{-1} [49] versus 21.6 kcal mol^{-1} for norbornene [50]; after cycloaddition, the cyclopropane product has a decreased strain of 28.7 kcal mol^{-1} [51]. Based on these considerations, we

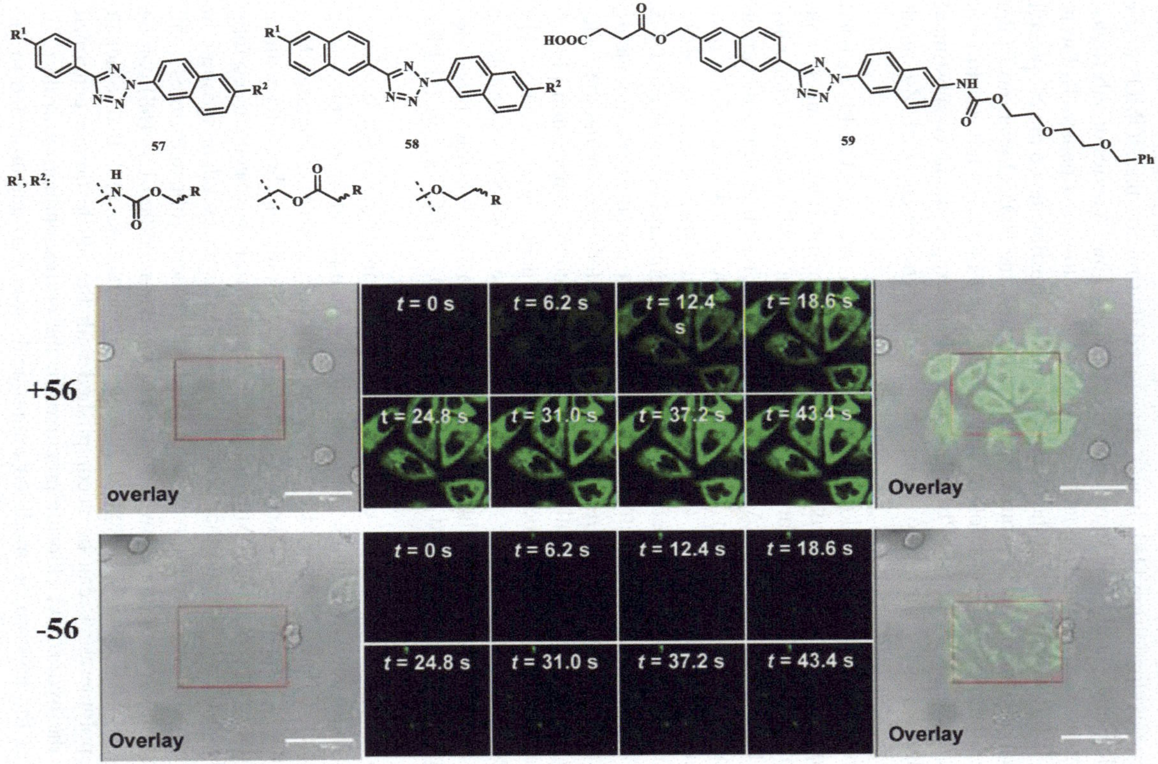

Fig. 2 Structure of 2PE photoactivatable tetrazoles (*top*) and fluorescent imaging of microtubules in CHO cells (*bottom*) using tetrazole **59** in the presence or in the absence of a fumarate-modified docetaxel **56**. The *rectangle area* were illuminated with the 700-nm femtosecond pulsed laser for the indicated time

synthesized a cyclopropene modified Lys, **CpK**, and showed that **CpK** can be site-specifically incorporated into proteins using the amber codon suppression strategy [52]. Fast cycloaddition kinetics was observed with the 3,3-disubstituated cyclopropene in 1:1 PBS/MeCN; the second-order rate constant was determined to be $58 \pm 16\ M^{-1}\ s^{-1}$ (for comparison: $k_2 = 46 \pm 9\ M^{-1}\ s^{-1}$ for acryl amide, $32 \pm 12\ M^{-1}\ s^{-1}$ for norbornene, and $0.95\ M^{-1}\ s^{-1}$ for allyl phenyl ether). Importantly, **CpK** was shown to direct site-specific modification of green fluorescent protein inside HEK293T cells via the tetrazole-based photoclick chemistry. Notably, the cyclopropene moiety has also been used as a reaction partner in tetrazine ligation [53, 54], with its reactivity preference between the tetrazine and tetrazole chemistries depending on the substitution pattern [55]. Realizing the cyclopropene reactivity can be further increased by reducing the steric hindrance at position 3, we fused a cyclobutane ring with the cyclopropene to generate spiro[2.3]hex-1-ene (**Sph**) [56]. Sph reacted with methoxy-diphenyltetrazole in CD_3CN 17-times faster than the 3,3-disubstitued cyclopropene. A spiro[2.3]hex-1-ene modified lysine (**SphK**) was synthesized and shown to be successfully incorporated into the superfolder GFP (sfGFP) using the wild-type PylRS/tRNA pair. To our delight, the **SphK**-encoded sfGFP displayed a fast reaction kinetics with a water-soluble tetrazole in phosphate buffer in the photo-triggered cycloaddition reaction with a measured k_2 value of $1.0 \times 10^4\ M^{-1}\ s^{-1}$, comparable to the typical tetrazine-*trans*-cyclooctene cycloaddition reaction ($k_2 = 2.2 \times 10^4\ M^{-1}\ s^{-1}$) [57].

On the other hand, Liu and co-workers demonstrated the accelerated photoclick chemistry with proteins using the electron-deficient acrylamide modified lysine (**AcrK**) [58]. The basis for faster kinetics with acrylamide is its lower LUMO energy compared to allyl phenyl ether. The **AcrK**-encoded sfGFP was successfully labeled with the hydrazonyl chloride precursor. The utility of AcrK was further demonstrated with fluorescent labeling of an overexpressed membrane protein OmpX in *E. coli.* Independently, Wang and co-workers evolved an orthogonal tRNA/aminoacyl-tRNA synthetase pair that allowed selective incorporation of **AcrK** into bacterial tubulin-like cytoskeleton protein FtsZ [59]. The utility of **AcrK** in the tetrazole-based photoclick chemistry was demonstrated with the purified proteins in vitro as well as in *E. coli* and mammalian cells. Additionally, **AcrK** was incorporated into GFP-TAG-mCherry-HA protein in *Arabidopsis thaliana*, a commonly used plant model.

The recent development of super-resolution microscopic techniques (PALM [60] and STORM [61]) has demanded new photoactivatable fluorophores with high turn-on efficiency and excellent biocompatibility. Realizing that the pyrazoline adducts formed from the photo-triggered cycloaddition reaction are fluorescent, we envisioned that a new type of fluorescent turn-on probes can be desired by combining the tetrazole and alkene groups on the same molecule [62]. The resulting alkene-modified tetrazoles (**69**) can be efficiently activated using a mild UV light source (302 or 365 nm), and the pyrazoline products (**70**) can be readily monitored under fluorescence microscope. We demonstrated this intramolecular cycloaddition reaction approach to turn-on probe design by imaging the microtubules in intact CHO cells treated with the alkene–tetrazole modified paclitaxel derivatives (**69**, Scheme 9). Similarly, the tetrazole-based intramolecular photoclick chemistry was also applied to synthesize the side chain-crosslinked peptides [63]. Briefly, the tetrazole and acrylamide modified lysine or ornithine were incorporated into a model 3_{10}-helical peptide [64, 65]. After photoirradiation, the pyrazoline-crosslinked peptides were generated in high yields. This tetrazole-based photo-crosslinking strategy was later successfully applied to the synthesis of cell-permeable peptide dual inhibitors of the p53-Mdm2/Mdmx interactions [66].

The development of smart hydrogels as extracellular matrices has shown growing importance in cell culture [67]. By taking advantage of the spatiotemporal control of the photo-responsible compounds, Zhong and co-workers used the photo-triggered

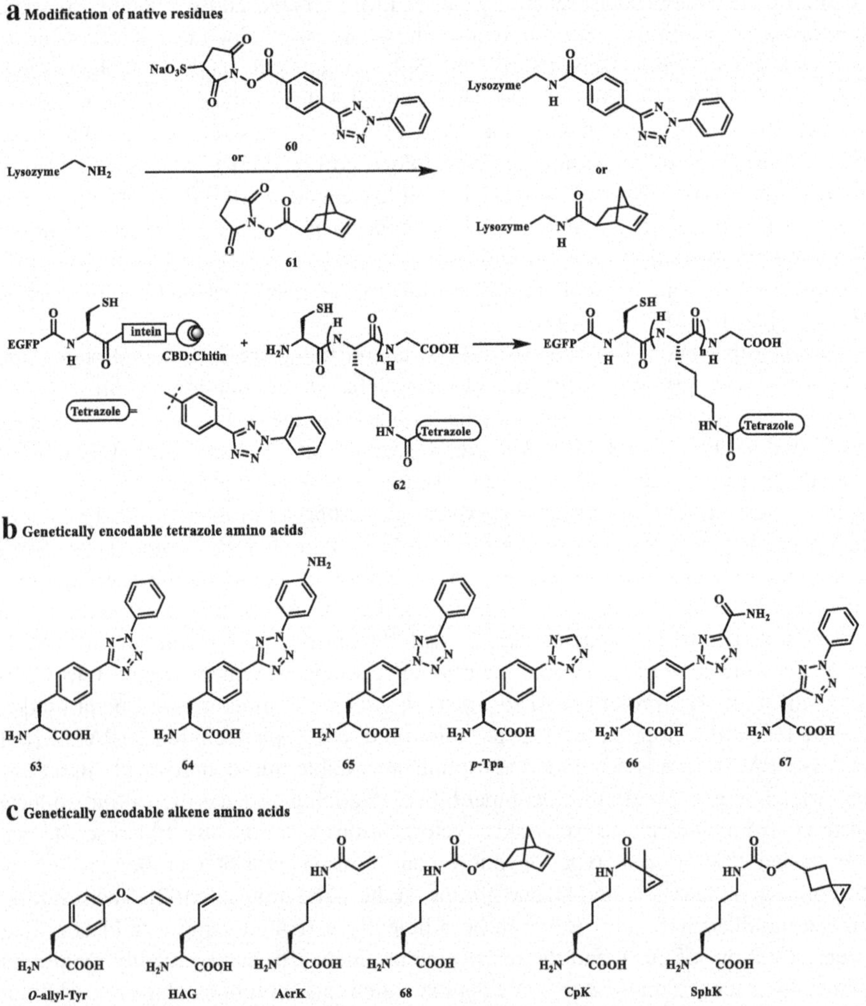

Scheme 8 Strategies for site selective protein modifications via photo-triggered tetrazole-alkene cycloaddition reaction: **a** modification of native residues, **b** genetic encoding of tetrazole amino acids, and **c** genetic encoding of alkene amino acids

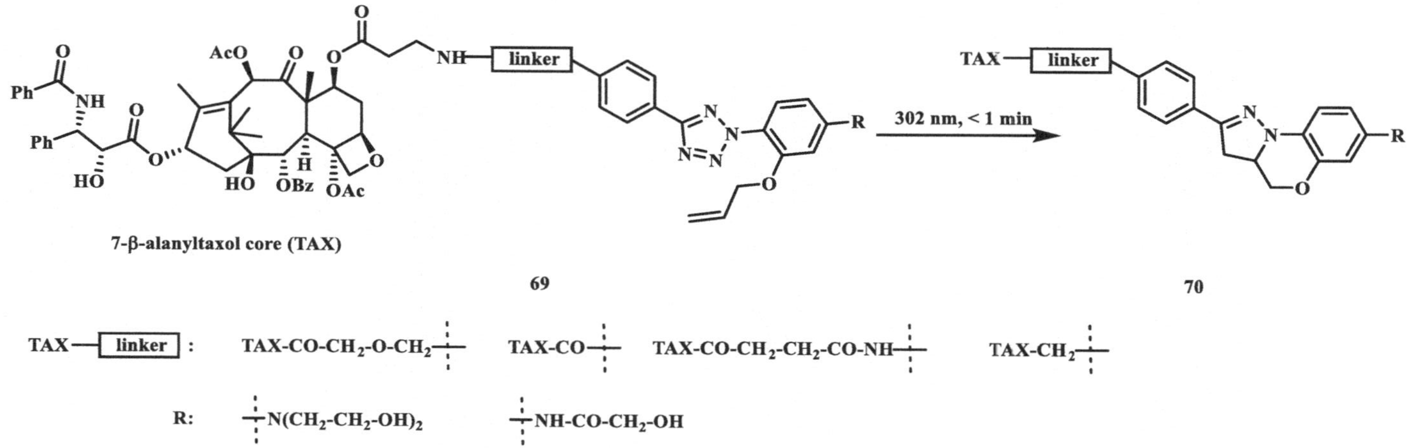

Scheme 9 Tetrazole-based turn-on fluorescent probes for imaging microtubules in live cells

tetrazole–methacrylate cycloaddition reaction to generate the hydrogels (Fig. 3a) [68]. This photo-triggered gelation approach offered a number of advantages, including tunable gelation by varying exposure time, high specificity and conversion, and avoidance of protein denaturation. Separately, Zhang and co-workers applied the intramolecular tetrazole-based photoclick chemistry to regulate the self-assembly of the hydrogels (Fig. 3b) [69]. Irradiation of the hydrogels induced degradation of the supramolecular matrix, which led to the release of encapsulated biological materials such as proteins and cells. For example, the photo-triggered release control of the encapsulated horse serum was shown, which induced the differentiation of C2C12 cells [70] on top of the hydrogels.

3 1,3-Dipolar Cycloaddition Reaction Between Azirines and Alkenes

Photo-triggered ring opening of 2*H*-azirines is a well-known reaction to produce pyrrolines [8, 71]. Padwa and co-workers showed that photoirradiation of azirines with a mercury arc lamp (450 W) equipped with Vycor filter generated the reactive nitrile ylide intermediate (**72**), which can be stabilized by the phenyl substituents. The nitrile ylide (**72**) then reacts with the electron-deficient olefins (**73**) such as acrylate and acrylonitrile in a cycloaddition reaction to form Δ^1-pyrrolines (**74**) (Scheme 10) [8]. Steenken and co-workers studied reaction kinetics of azirines with dipolarophiles as well as nucleophiles such as alcohols [72]. They showed that the reaction rate depends on the azirine substituents, the nucleophilicity of the reactant and the acidity of the alcohol.

In 2010, we reported the first use of photo-triggered azirines-mediated cycloaddition reaction for selective protein modification in biological buffer (Scheme 11) [73]. A series of azirines were synthesized and their reactions with 2-methylmethacrylate and dimethyl fumarate in EtOH/H_2O (1:1) were investigated. The photo-generated nitrile ylide intermediate was not detected in the HPLC analysis, suggesting that the rate-determining step is likely the azirine ring opening with a measured first-order rate constant of 0.0379 s^{-1}. Furthermore, an azirine-modified lysozyme was selectively functionalized by a water-soluble PEGylated fumarate (**78**) after 2 min 302-nm photoirradiation. The major limitation of this reaction is the need of highly electron-deficient dipolarophiles such as fumarate.

Very recently, Blinco and Barner-Kowollik have successfully shifted the photoactivation wavelength into the visible region (>390 nm) [74]. By replacing the phenyl group with the pyrene group on the azirine ring, they showed that the photo-triggered cycloaddition proceeded in <1 min under ambient conditions with the electron-deficient alkenes such as fumarates, maleimide, acrylates, and activated acetylenes. Reactions were initiated using a photoreactor equipped with an LED light (410–420 nm, 3 W). The utility of this modified reaction was demonstrated through efficient conjugation of PEG chains to the pyrene fluorophore.

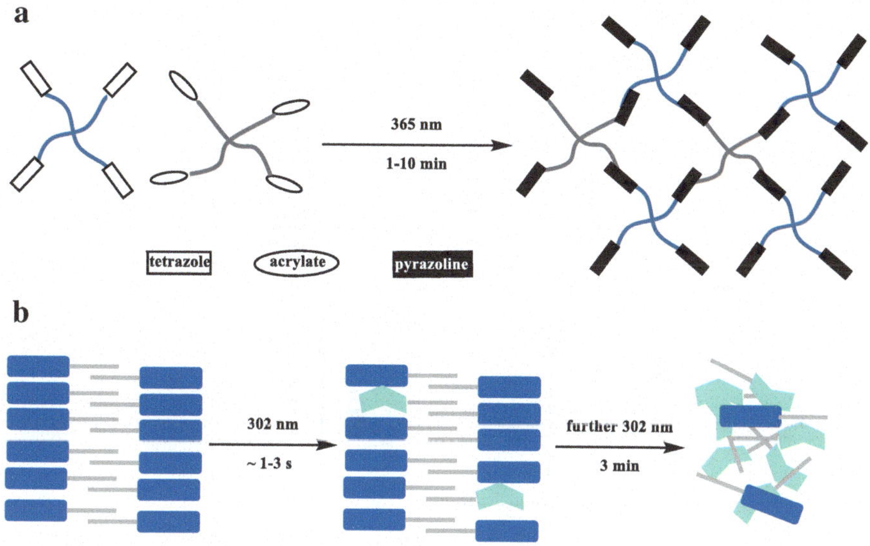

Fig. 3 **a** Hydrogel formation with the tetrazole-based photoclick reaction. **b** The collapse of hydrogels induced by intramolecular photoclick reaction

4 Photoinduced Hetero Diels–Alder Reactions

Diels–Alder (DA) reaction is one of the most useful chemical transformations. This reaction typically proceeds slowly in the absence of a catalyst, but can be accelerated with the use of Lewis acids [75]. It is characterized by high yield under various conditions, has less sensitivity against solvents, and does not produce byproducts. It is characterized by high yield under various conditions, insensitive toward solvent polarity, and little byproducts. Indeed, the tetrazine/*trans*-cyclooctene based inverse electron-demand DA reaction has become the most popular bioorthogonal reaction in the literature because of its fast reaction kinetics [76]—second-order rate constant as high as $2.8 \times 10^6\ M^{-1}\ s^{-1}$ has been reported [77]. So far, two photo-triggered hetero-dienes have been exploited in the DA reactions for biological use, namely, *o*-quinone methides and hydroxy-*o*-quinodimethanes.

4.1 Naphthoquinone Methide

In 2011, Popik and co-workers reported photochemical dehydration of 3-hydroxy-2-naphthalenemethanol (*o*-naphthoquinone precursor, NQMP, **79**) derivatives to *o*-naphthoquinone methides (*o*NQMs, **80**) [78, 79]. The in situ generated reactive intermediate *o*NQM underwent facile cycloaddition with vinyl ethers (**81**) to form photostable benzochromans (**82**) (Scheme 12). NQMP (**79**) has two major absorption bands at 275 nm ($\log \varepsilon = 4.06$) and 324 nm ($\log \varepsilon = 3.70$); thus it can be photoactivated with either low pressure mercury lamp (254 nm) or fluorescent tubes (300 and 350 nm). The quantum yield of the photoactivation was very high ($\Phi_{300} = 0.17 \pm 0.02$ for **79**). The in situ generated *o*NQMs was quenched mostly by

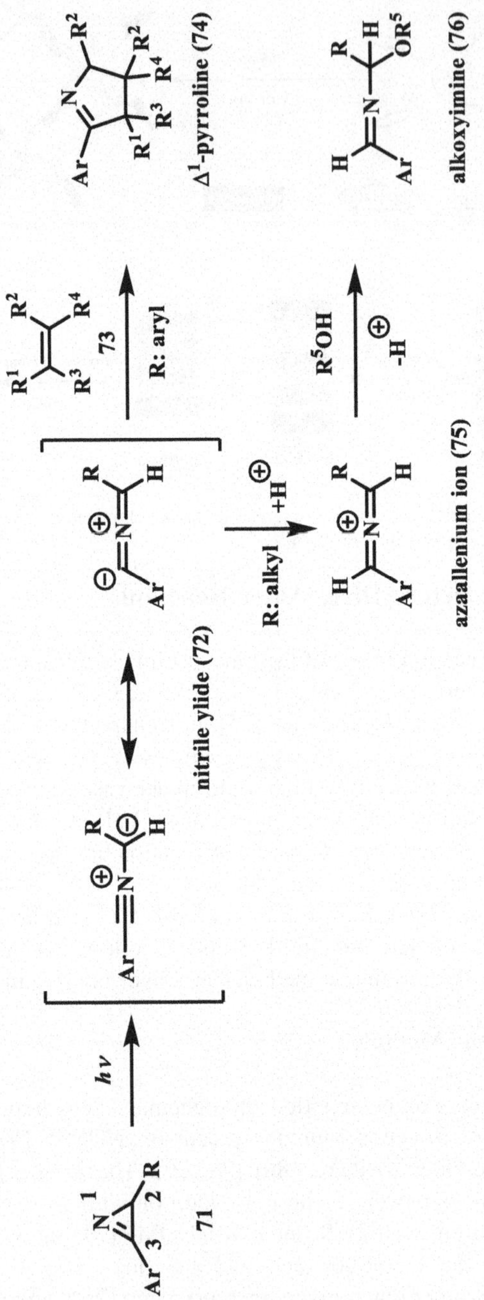

Scheme 10 Photo-triggered reactions of azirines and alkenes or alcohols

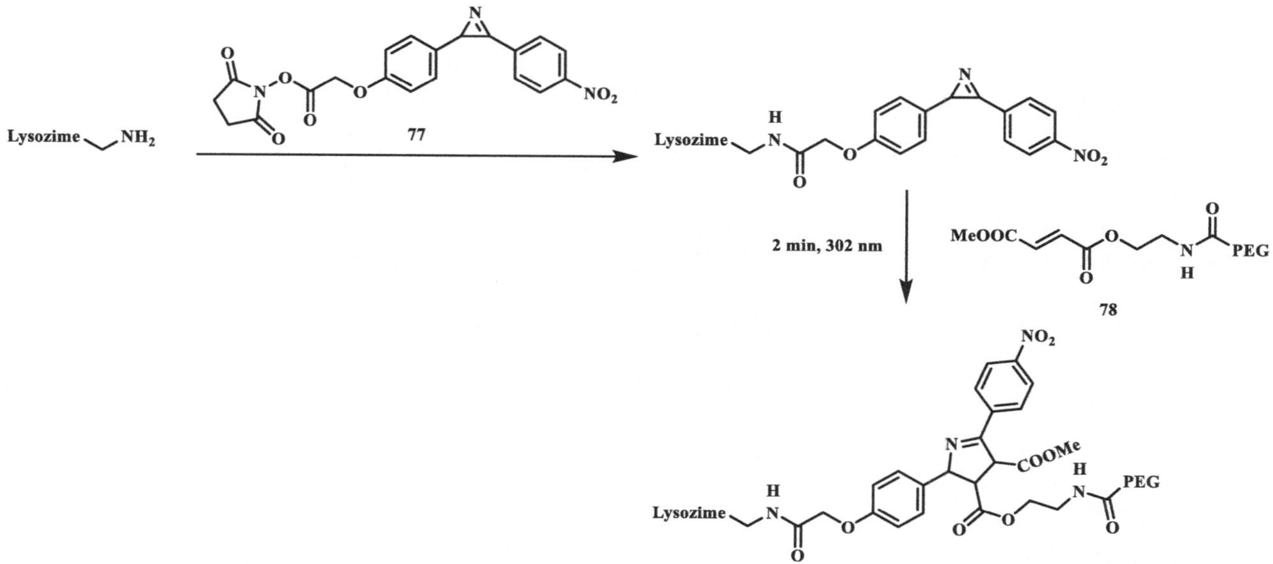

Scheme 11 Selective modification of an azirine-containing lysozyme by a PEG-modified fumarate via a photo-triggered cycloaddition reaction

water with a high reaction rate ($k_w = 145\ s^{-1}$ or $2.61\ M^{-1}\ s^{-1}$), which regenerates the starting material. When vinyl ethers was present, *o*NQMs underwent rapid cycloaddition reaction with a second order rate constant in the range of $4–6 \times 10^4\ M^{-1}\ s^{-1}$ in PB/MeCN (1:1). For photoirradiation, samples were exposed to 4-W 300-nm fluorescent tubes for 3 min (~15 % conversion) or 20 min (>99 % conversion). Notably, 350-nm fluorescent tubes also worked with longer exposure time, presumably due to the lower absorption at this wavelength.

When other alkenes such as methyl acrylate, dimethyl maleate, 2,5-dihydrofurans and methyl cyclohexene were present in a competition reaction, only alkyl vinyl ethers produced the desired benzochroman cycloadducts (**84**) (Scheme 13). Mechanistic studies revealed that the key issue is the competition with hydration. Apparently, the cycloaddition reactions with the non-polarized electron-deficient alkenes could not compete with the water-quenching.

Since acetals are susceptible to hydrolysis, the adjustment of solution pH offers a control over the product stability; at neutral or basic pH, no hydrolysis would occur. The enamines (**86**) are another class of electron-rich polarized alkenes which could undergo facile cycloaddition reaction with *o*NQMs (**80**), but the primary cycloadducts (**88**) rapidly hydrolyze in aqueous medium to produce 2-hydroxy-benzochromanes (**90**). Nevertheless, this method is suitable for the introduction of functional groups at position 3 of the benzochromane (**90**) (Scheme 14).

Reactive *o*NQMs (**80**) can also be efficiently captured by strong nucleophiles such as thiols presented in millimolar concentrations in biological systems. They observed thiol addition ($k_2 = 2.2 \times 10^5\ M^{-1}\ s^{-1}$ with thioethanolamine; 5-times faster than the DA reaction with ethyl vinyl ether). However, the adduct was photochemically labile; upon longer irradiation time (30 min instead of 20 min) it was converted back to NQMP **79**) (Scheme 14). The feature that oNQMs react faster with thiols (in reversible way) but slower than vinyl ethers (in irreversible way) was exploited in the controlled release of thiol functionalized compounds *o*NQMs (Scheme 15) [80]. First, with a short irradiation (2–5 min, 300 or 350 nm) the thiol adduct (**92**) was formed, which was later released quantitatively with 2–5 min (350 nm) irradiation in the presence of ethyl vinyl ether. Selective and reversible functionalization of proteins was achieved through Cys modification. Reversibility was also shown with photoinitiated capture and hydrolytic cleavage (Scheme 15) [81].

Lei and coworkers modified this structure to allow a reaction with vinyl thioethers (VT) to form a stable covalent bond [82]. The biological utility was

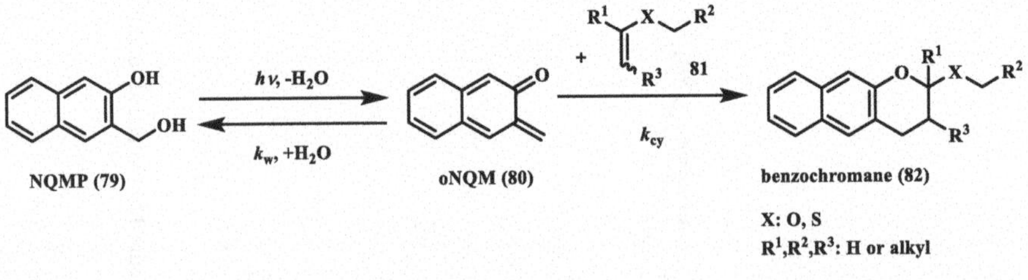

Scheme 12 Photoactivation of NQMP to generate *o*NQM, which then reacts with vinyl (thio)ether

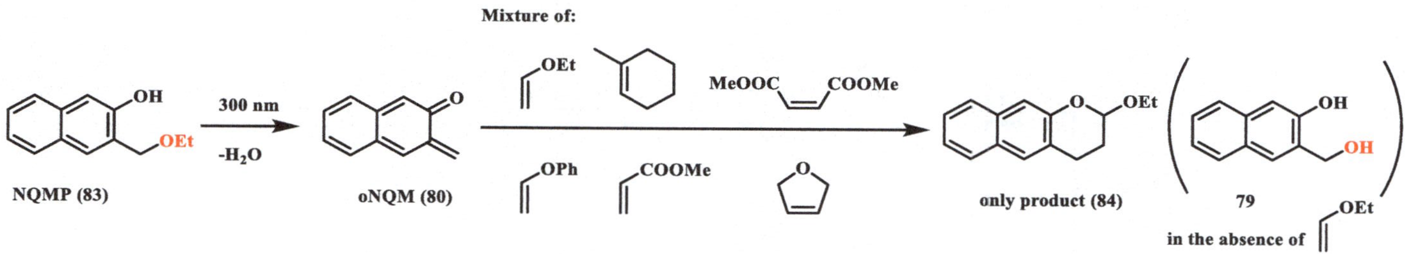

Scheme 13 Substrate scope of the alkene dienophiles in the photo-triggered cycloaddition reaction; product was formed only with ethyl vinyl ether

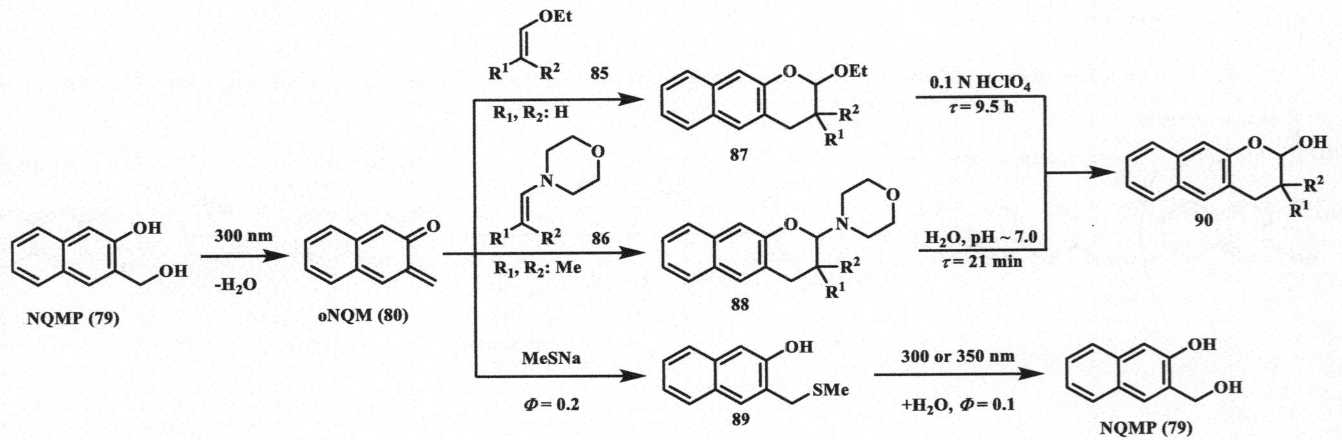

Scheme 14 Reactions between NQM and the electron-rich alkenes or thiol

demonstrated by labeling a VT-modified BSA and imaging a VT-modified taxol in live HeLa cells. Introducing a N into the aromatic system eliminate the need for irradiation, therefore losing the advantage provided by photoinitiation.

4.2 *o*-Quinodimethane

Photoenolization of *ortho*-methyl phenyl ketones and aldehydes (**96**) has been used in organic chemistry for a long time [6, 83, 84]. The first application of this reaction in polymer science was made by Barner-Kowollik et al. [85]. Irradiation leads to formation of biradicals (**99**) that rearrange to highly reactive diene (hydroxy-*o*-quinodimethane or "photoenol", **100**). This intermediate **100** reacts with electron-deficient dienophiles (**101**) such as maleimide via Diels–Alder reaction (Scheme 16). The driving force is the restored aromaticity. In their study, two polymers were coupled through maleimide and *o*-methyl phenyl ketone functional groups irradiated with compact fluorescence lamp (36 W, $\lambda_{max} = 320$ nm) in toluene for 20–100 min at ambient temperature. Triblockpolymers were synthesized using the same strategy [86].

Later, the same group modified the structure of *o*-methyl phenyl ketone to 2-formyl-3-methylphenoxy (FMP, **103**) and reached higher reactivity towards the alkene dienophiles (Scheme 17) [87]. They assumed that after photoexcitation the hydrogen bond formation stabilizes the reactive intermediate (**104**) by increasing both its lifetime and ratio of formed *Z* isomer, which has higher reactivity towards dienophiles relative to the *E* isomer [88]. The less electron-deficient acrylates, which cannot react with the original hydroxy-*o*-quinodimethane, can now be used with FMP [89]. Full conversion with maleimide was reached in <15 min in acetonitrile and dichloromethane. While the reaction is water compatible, long time is needed when the reactions are performed in polar solvents such as water and DMF in order to achieve full conversion. For biological applications, a preliminary study was carried out with the maleimide-oligopeptide conjugate (**105**). The product was formed in quantitative yield from the peptide and the FMP (**103**) in PBS/MeCN

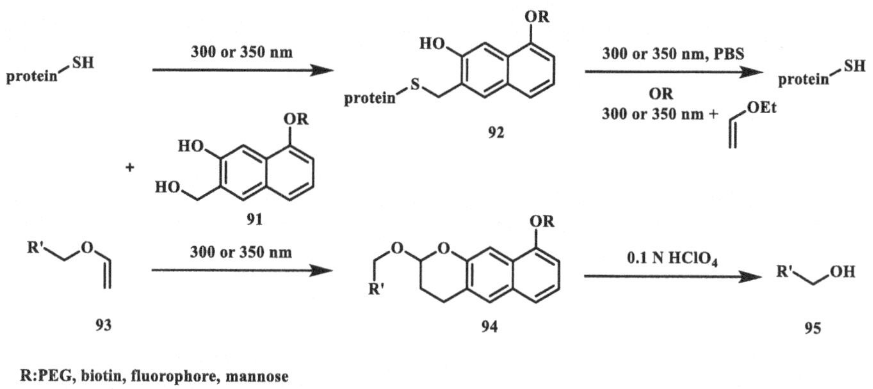

Scheme 15 Exploiting the reversibility: capture with *o*NMQs and release through photolysis or hydrolysis

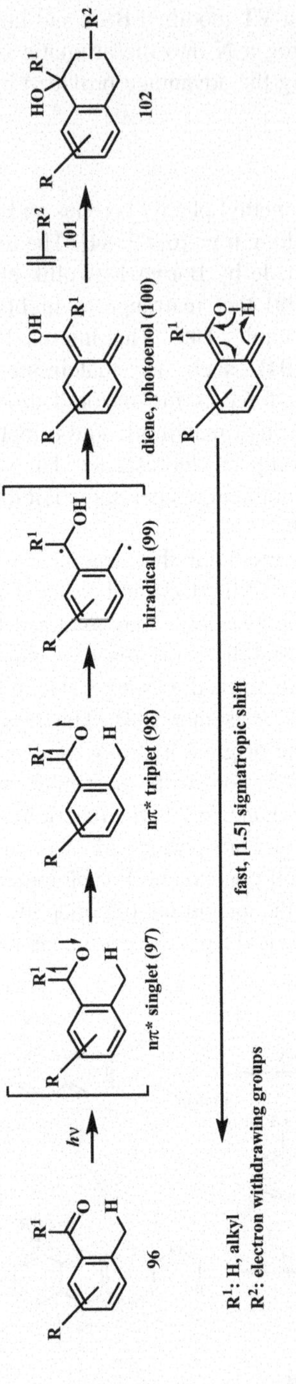

Scheme 16 Mechanism of the photo-triggered isomerization and hetero Diels–Alder reaction of *o*-methyl phenyl ketones and aldehydes

(1:3) under 320 nm irradiation. Full conversion was reached after 2 h photoirradiation. Patterning of silicon surfaces with oligopeptides (GRGSGR) was also demonstrated.

5 Strain-Promoted Azide–Alkyne Cycloaddition

Copper-catalyzed azide–alkyne cycloaddition (CuAAC) is one of the most powerful click reactions. The only disadvantage is that the copper is toxic to certain cells [90]. Despite efforts to make the copper complexes more biocompatible [91, 92], the breakthrough was achieved by the Bertozzi group [93] through harnessing the ring strain present in cyclooctyne to accelerate the reaction. A variety of cyclooctynes and one cycloheptyne have subsequently been reported [94, 95].

Based on the earlier studies of using cyclopropenone (CP) [9, 96] as a photo-protected precursor of alkyne, Popik and co-workers combined this feature with the click chemistry (**109–110**) (Scheme 18) [97]. The CP is thermally stable in aqueous solution and reacts with azides exclusively after photoirradiation. High quantum yield was measured for the photo-decarbonylation step ($\Phi_{355} = 0.33$) and the corresponding alkyne (**111–112**) was obtained in quantitative yield. The cycloaddition rate was determined to be 1.6–7.6 $\times$ 10^{-2} M^{-1} s^{-1} in MeOH. For biological applications, Jurkat cells were cultured in the presence of peracetylated *N*-azidoacetyl mannosamine ($Ac_4ManNAz$). After incubation with a biotin-modified CP (**113**) and 1 min irradiation at 350 nm, the labeling efficiency was evaluated with avidin-fluorescein isothiocyanate (FITC). The labeling efficiencies were comparable between the in situ formed cyclooctyne and the unprotected one. Without photoirradiation, the background labeling was negligible. The temporal control could not be demonstrated because of 30-min delay between the irradiation and fluorescence acquisition. One disadvantage of CP reagent is that it is prone to nucleophilic attack by endogenous thiols [9].

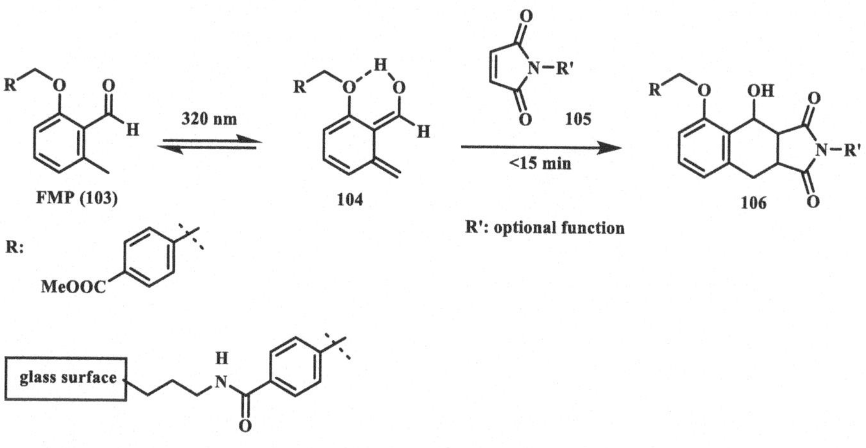

Scheme 17 Photo-triggered hetero Diels–Alder reaction with FMP

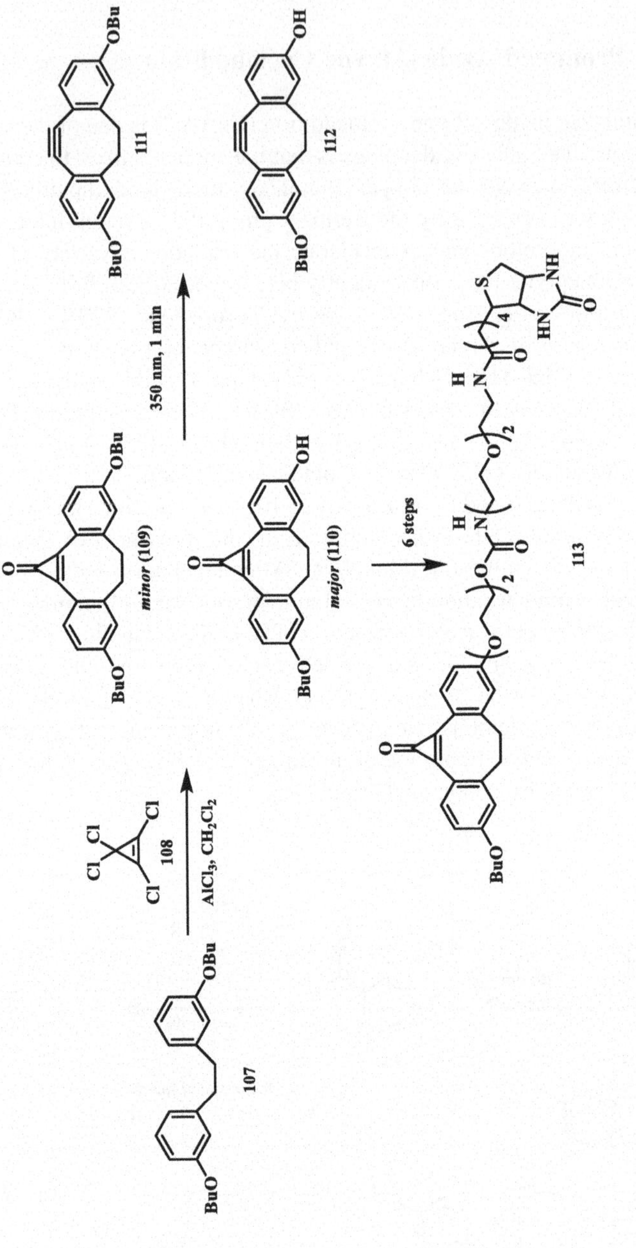

Scheme 18 Synthesis and photodecarbonylation of dibenzocyclopropenones

Table 1 Optimized parameters for the photo-triggered click reactions

Cycloaddition reaction	Irradiation wavelength	Irradiation time	Rate constant (k_2) (solvent)
Tetrazole–alkene	405, 700 nm (2PE)	~1 min	3.4×10^4 M^{-1} s^{-1} (1:1 PB/MeCN)
Azirine–alkene	302 nm	2 min	0.0379 M^{-1} s^{-1} (1:1 PBS/MeCN)
3-Hydroxy-2-naphthalenemethanol-vinyl ether	300, 350 nm	3–20 min	4–6 $\times 10^4$ M^{-1} s^{-1} (1:1 PB:MeCN)
o-Methyl phenyl ketone/aldehyde-alkene	320 nm	2 h	Not determined
Cyclopropenone-azide	350 nm	1 min	7.6×10^{-2} M^{-1} s^{-1} (MeOH)

6 General Conclusions

For a long time, photo-triggered click chemistry has been used mostly in material sciences. With the recent design and synthesis of some new reagents, these reactions have begun to be employed in bioorthogonal chemistry based studies. The advantages of light-induced spatiotemporal control has been harnessed for selective labeling of the biomolecules in living system. Table 1 summarizes the optimized parameters achieved so far with the best reactant pairs for each of these photo-triggered cycloaddition reactions. In conjunction of the recent rapid development of synthetic biology in which the unique chemical groups are added to cell's basic building blocks, we anticipate the photo-triggered bioorthogonal reactions will offer an unprecedented spatiotemporally controlled chemical tool to dissect complex biological processes in their native environments.

Acknowledgments Work on the tetrazole-based photoclick chemistry in QL lab was supported by the National Institutes of Health (GM 085092). AH thanks the Rosztoczy Foundation (to A.H.) for a scholarship.

References

1. Kolb HC, Finn MG, Sharpless KB (2001) Angew Chem Int Ed 40:2004
2. Saxon E, Bertozzi CR (2000) Science 287:2007
3. Prescher JA, Bertozzi CR (2005) Nat Chem Biol 1:13
4. Lowe AB (2014) Polym Chem 5:4820
5. Lowe AB (2014) Polymer 55:5517
6. Sammes PG (1976) Tetrahedron 32:405
7. Diao L, Yang C, Wan P (1995) J Am Chem Soc 117:5369
8. Padwa A (1976) Acc Chem Res 9:371
9. Poloukhtine A, Popik VV (2003) J Org Chem 68:7833
10. Mayer G, Heckel A (2006) Angew Chem Int Ed 45:4900
11. Su MD, Liao HY, Chung WS, Chu SY (1999) J Org Chem 64:6710
12. Huisgen R, Seidel M, Wallbillich G, Knupfer H (1962) Tetrahedron 17:3
13. Clovis JS, Eckell A, Huisgen R, Sustmann R (1967) Chem Ber 100:60
14. Toubro NH, Holm A (1980) J Am Chem Soc 102:2093

15. Zheng SL, Wang Y, Yu Z, Lin Q, Coppens P (2009) J Am Chem Soc 131:18036
16. Weinberg P, Csongar C, Grummt UW (1989) J Photochem Photobiol A 50:11
17. Lohse V, Leihkauf P, Csongar C, Tomaschewski G (1988) J Prakt Chem 330:406
18. Houk KN, Sims J, Watts CR, Luskus LJ (1973) J Am Chem Soc 95:7301
19. Molteni G, Orlandi M, Broggini G (2000) J Chem Soc Perk Trans 1:3742
20. Padwa A, Nahm S, Sato E (1978) J Org Chem 43:1664
21. Meier H, Heimgartner H (1985) Helv Chim Acta 68:1283
22. Darkow R, Yoshikawa M, Kitao T, Tomaschewski G, Schellenberg J (1994) J Polym Sci A Polym Chem 32:1657
23. Kaya E, Vrabel M, Deiml C, Prill S, Fluxa VS, Carell T (2012) Angew Chem Int Ed 51:4466
24. Wang XS, Lee YJ, Liu WR (2014) Chem Commun 50:3176
25. Ito S, Tanaka Y, Kakehi A, Kondo K (1976) Bull Chem Soc Jpn 49:1920
26. Koguro K, Oga T, Mitsui S, Orita R (1998) Synthesis 6:910
27. Lenda F, Guenoun F, Tazi B, Larbi NB, Allouchi H, Martinez J, Lamaty F (2005) Eur J Org Chem 326
28. An P, Yu Z, Lin Q (2013) Chem Commun 49:9920
29. Wang Y, Rivera Vera CI, Lin Q (2007) Org Lett 9:4155
30. Wells RL, Han A (1985) Int J Radiat Biol 47:17
31. Wang Y, Hu WJ, Song W, Lim RKV, Lin Q (2008) Org Lett 10:3725
32. Yu Z, Ho LY, Wang Z, Lin Q (2011) Bioorg Med Chem Lett 21:5033
33. An P, Yu Z, Lin Q (2013) Org Lett 15:5496
34. Yu Z, Ohulchanskyy TY, An P, Prasad PN, Lin Q (2013) J Am Chem Soc 135:16766
35. McClure DS (1954) J Chem Phys 22:1668
36. Furuta T, Wang SS, Dantzker JL, Dore TM, Bybee WJ, Callaway EM, Denk W, Tsien RY (1999) Proc Natl Acad Sci USA 96:1193
37. Lang K, Chin JW (2014) ACS Chem Biol 9:16
38. Liu CC, Schultz PG (2010) Annu Rev Biochem 79:413
39. Song W, Wang Y, Qu J, Madden MM, Lin Q (2008) Angew Chem Int Ed 47:2832
40. Song W, Yu Z, Madden MM, Lin Q (2010) Mol BioSyst 6:1576
41. Song W, Wang Y, Qu J, Lin Q (2008) J Am Chem Soc 130:9654
42. Wang Y, Song W, Hu WJ, Lin Q (2009) Angew Chem Int Ed 48:5330
43. Wang Y, Lin Q (2009) Org Lett 11:3570
44. Wang J, Zhang W, Song W, Wang Y, Yu Z, Li J, Wu M, Wang L, Zang J, Lin Q (2010) J Am Chem Soc 132:14812
45. Yu Z, Lim RKV, Lin Q (2010) Chem Eur J 16:13325
46. Lang K, Davis L, Torres-Kolbus J, Chou C, Deiters A, Chin JW (2012) Nat Chem 4:298
47. Plass T, Milles S, Koehler C, Szymański J, Mueller R, Wießler M, Schultz C, Lemke EA (2012) Angew Chem Int Ed 124:4166
48. Bianco A, Townsley FM, Greiss S, Lang K, Chin JW (2012) Nat Chem Biol 8:748
49. Bach RD, Dmitrenko O (2004) J Am Chem Soc 126:4444
50. Schleyer PVR, William JE, Blanchard KR (1970) J Am Chem Soc 92:2377
51. Gordon MS (1980) J Am Chem Soc 102:7419
52. Yu Z, Pan Y, Wang Z, Wang J, Lin Q (2012) Angew Chem Int Ed 51:10600
53. Patterson DM, Nazarova LA, Xie B, Kamber DN, Prescher JA (2012) J Am Chem Soc 134:18638
54. Yang J, Seckute J, Cole CM, Devaraj NK (2012) Angew Chem Int Ed 51:7476
55. Kamber DN, Nazarova LA, Liang Y, Lopez SA, Patterson DM, Shih HW, Houk KN, Prescher JA (2013) J Am Chem Soc 135:13680
56. Yu Z, Lin Q (2014) J Am Chem Soc 136:4153
57. Taylor MT, Blackman ML, Dmitrenko O, Fox M (2011) J Am Chem Soc 133:9646
58. Lee Y-J, Wu B, Raymond JE, Zeng Y, Fang X, Wooley KL, Liu WR (2013) ACS Chem Biol 8:1664
59. Li F, Zhang H, Sun Y, Pan Y, Zhou J, Wang J (2013) Angew Chem Int Ed 52:9700
60. Betzig E, Patterson GH, Sougrat R, Lindwasser OW, Olenych S, Bonifacino JS, Davidson MW, Lippincott-Schwartz J, Hess HF (2006) Science 313:1642
61. Rust MJ, Bates M, Zhuang X (2006) Nat Methods 3:793
62. Yu Z, Ho LY, Lin Q (2011) J Am Chem Soc 133:11912
63. Madden MM, Rivera Vera CI, Song W, Lin Q (2009) Chem Commun 5588
64. Karle IL, Flippen-Anderson JL, Uma K, Balaram P (1990) Proteins 7:62
65. Karle IL, Flippen-Anderson JL, Uma K, Balaram P (1993) Biopolymers 33:827

66. Madden MM, Muppidi A, Li Z, Li X, Chen J, Lin Q (2011) Bioorg Med Chem Lett 21:1472
67. Alge DL, Anseth KS (2013) Nat Mater 12:950
68. Fan Y, Deng C, Cheng R, Meng F, Zhong Z (2013) Biomacromolecules 14:2814
69. He M, Li J, Tan S, Wang R, Zhang Y (2013) J Am Chem Soc 135:18718
70. Simone C, Forcales SV, Hill DA, Imbalzano AN, Latella L, Puri PL (2004) Nat Genet 36:738
71. Padwa A, Smolanoff J (1971) J Am Chem Soc 93:548
72. Albrecht E, Mattay J, Steenken S (1997) J Am Chem Soc 119:11605
73. Lim RKV, Lin Q (2010) Chem Commun 46:7993
74. Mueller JO, Schmidt FG, Blinco JP, Barner-Kowollik C (2015) Angew Chem Int Ed 54:10284
75. Carey FA, Sundberg RJ (2007) Advanced organic chemistry: part b: reactions and synthesis, 5th edn. Springer, New York, pp 481–487. E-ISBN 978-0-387-44899-2 (part B)
76. Liu F, Paton RS, Kim S, Liang Y, Houk KN (2013) J Am Chem Soc 135:15642
77. Rossin R, van den Bosch SM, ten Hoeve W, Carvelli M, Versteegen RM, Lub J, Robillard MS (2013) Bioconj Chem 24:1210
78. Arumugam S, Popik VV (2011) J Am Chem Soc 133:15730
79. Arumugam S, Orski SV, Locklin J, Popik VV (2012) J Am Chem Soc 134:179
80. Arumugam S, Guo J, Mbua NE, Friscourt F, Lin N, Nekongo E, Boons GJ, Popik VV (2014) Chem Sci 5:1591
81. Arumugam S, Popik VV (2014) J Org Chem 79:2702
82. Li Q, Dong T, Liu X, Lei X (2013) J Am Chem Soc 135:4996
83. Charlton JL, Alauddin MM (1987) Tetrahedron 43:2873
84. Segura JL, Martin N (1999) Chem Rev 99:3199
85. Gruendling T, Oehlenschlaeger KK, Frick E, Glassner M, Schmid C, Barner-Kowollik C (2011) Macromol Rapid Commun 32:807
86. Glassner M, Oehlenschlaeger KK, Gruendling T, Barner-Kowollik C (2011) Macromolecules 44:4681
87. Pauloehrl T, Delaittre G, Winkler V, Welle A, Bruns M, Boerner HG, Greiner AM, Bastmeyer M, Barner-Kowollik C (2012) Angew Chem Int Ed 51:1071
88. Mellows SM, Sammes PG (1971) J Chem Soc Chem Comm 21
89. Winkler M, Mueller JO, Oehlenschlaeger KK, de Espinosa LM, Meier MAR, Barner-Kowollik C (2012) Macromolecules 45:5012
90. Kennedy DC, McKay CS, Legault MCB, Danielson DC, Blake JA, Pegoraro AF, Stolow A, Mester Z, Pezacki JP (2011) J Am Chem Soc 133:17993
91. Haldón E, Nicasio MC, Pérez PJ (2015) Org Biomol Chem 13:9528
92. Yang M, Jalloh AS, Wei W, Zhao J, Wu P, Chen PR (2014) Nat Commun 5:4981
93. Agard NJ, Prescher JA, Bertozzi CR (2004) J Am Chem Soc 126:15046
94. Dommerholt J, van Rooijen O, Borrmann A, Guerra CF, Bickelhaupt FM, van Delft FL (2014) Nat Commun 5:5378
95. Gold B, Dudley GB, Alabugin IV (2013) J Am Chem Soc 135:1558
96. Poloukhtine A, Popik VV (2006) J Phys Chem A 110:1749
97. Poloukhtine AA, Mbua NE, Wolfert MA, Boons GJ, Popik VV (2009) J Am Chem Soc 131:15769

Top Curr Chem (Z) (2016) 374:3
DOI 10.1007/s41061-015-0005-z

REVIEW

Inverse Electron-Demand Diels–Alder Bioorthogonal Reactions

Haoxing Wu[1] · **Neal K. Devaraj**[1]

Received: 23 November 2015 / Accepted: 1 December 2015 / Published online: 22 December 2015

Abstract Bioorthogonal reactions have been widely used over the last 10 years for imaging, detection, diagnostics, drug delivery, and biomaterials. Tetrazine reactions are a recently developed class of inverse electron-demand Diels–Alder reactions used in bioorthogonal applications. Given their rapid tunable reaction rate and highly fluorogenic properties, tetrazine bioorthogonal reactions have come to be considered highly attractive tools for elucidating biological functions and messages in vitro and in vivo. In this chapter, we present recent advances expanding the scope of precursor reactivity and we introduce new biomedical methodology based on bioorthogonal tetrazine chemistry. We specifically highlight novel applications for different kinds of biomolecules, including nucleic acid, protein, antibodies, lipids, glycans, and bioactive small molecules, in the areas of imaging, detection, and diagnostics. We also briefly present other recently developed inverse electron-demand Diels–Alder bioorthogonal reactions. Lastly, we consider future directions and potential roles that inverse electron-demand Diels–Alder reactions may play in the fields of bioorthogonal and biomedical chemistry.

Keywords Diels–Alder · Tetrazine · Bioorthogonal · Imaging · Detection

1 Introduction

Bioorthogonal chemistry, as conceptualized by Bertozzi earlier this century, refers to chemical reactions that can take place within a physiological milieu with sufficient kinetics in a selective manner. Specifically, the reactants must be inert to

✉ Neal K. Devaraj
ndevaraj@ucsd.edu

1 Department of Chemistry and Biochemistry, University of California, San Diego, 9500 Gilman Drive, Urey Hall 4120, La Jolla, CA 92093, USA

all functional groups found in biological systems and reactive only with corresponding reaction partners. Furthermore, the reaction must not disturb any native biochemical processes [1–4]. Several classes of bioorthogonal chemistries have been developed, including strain-promoted alkyne-azide cycloaddition (SPAAC) [5–9], Staudinger ligation [10, 11], and 1,3-dipole cycloaddition [12–19], as well as Diels–Alder reaction [20–22]. With the improvement of reaction properties, bioorthogonal reactions have served as a versatile platform for selective labeling and manipulation of biomolecules in living systems [1, 23–26]. Thus, disease mechanisms and physiological processes can be elucidated at the molecular level. Numerous exciting applications in imaging, diagnostics, therapeutics, and biomaterials have been explored in recent years.

The Diels–Alder reaction is defined as a [4+2] cycloaddition between a conjugated diene and a substituted dienophile (alkene or alkyne) to form a (hetero-)cyclohexene system. Based on the electronic effects of the substituent on the diene and dienophile, Diels–Alder reactions can be classified as normal electron-demand (electron-rich diene reacts with electron-deficient dienophile) or inverse electron-demand (iEDDA, electron-deficient diene reacts with electron-rich dienophile) reactions (Scheme 1a). In a normal electron-demand Diels–Alder reaction, the electron-deficient dienophile, typically a Michael acceptor, is likely to be attacked by endogenous nucleophiles such as free amino and thiol groups in vivo. For this reason, the use of this reaction in bioorthogonal chemistry applications poses a challenge.

The most commonly used Diels–Alder bioorthogonal reactions are iEDDA. Tetrazine ligations, the most representative within the Diels–Alder bioorthogonal family, were first reported independently by two groups in 2008 [20, 21]. As shown in Scheme 1b, s-tetrazine and a strained alkene or alkyne derivative undergoes an inverse electron-demand Diels–Alder cycloaddition followed by a retro-Diels–Alder reaction to give the dihydropyridazine product and release N_2. Most importantly, this reaction proceeds rapidly at room temperature in aqueous environments without a catalyst, it is high-yielding, and the reactive partners are unreactive and compatible with a variety of functional groups and challenging conditions.

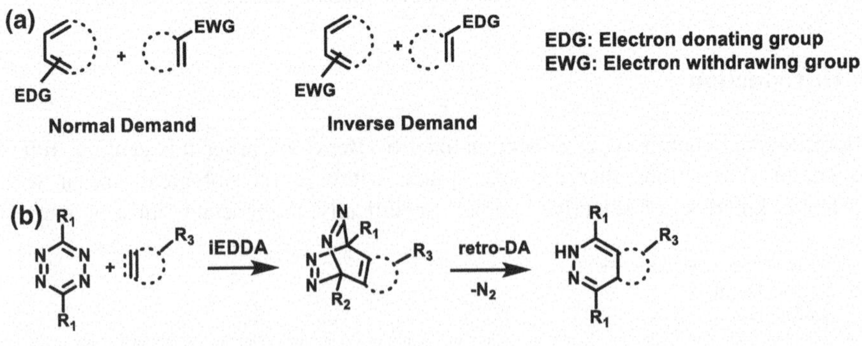

Scheme 1 **a** Normal electron-demand and inverse electron-demand Diels–Alder reaction. **b** Tetrazine bioorthogonal reaction. The reaction product is pyridazine if the dienophile is an alkyne

Moreover, the reaction rate is highly tunable across many orders of magnitude by chemically manipulating the electron deficiency of the tetrazine's 3,6 position substituents, or by employing different dienophiles, enabling the reaction class to meet the needs of many different applications. Even more impressive, tetrazine bioorthogonal reactions have fluorogenic properties when conjugated with certain fluorophores, facilitating highly useful applications in bimolecular imaging. For these reasons, tetrazine bioorthogonal reactions are attracting significant interest across multiple biomedical technologies. Several reviews have been published within the past few years with different areas of focus: click chemistry properties, biomedical applications, nanoparticle diagnostics, and tunability of dienophile properties [22, 23, 27–30]. Here, we introduce new chemistries of tetrazine as well as other iEDDA bioorthogonal strategies and summarize recent progress with various kinds of biomolecules.

2 Development of Inverse Electron-Demand Diels–Alder (iEDDA) Bioorthogonal Reactions

2.1 Recent Developments in Tetrazine Synthesis

The classical tetrazine synthetic methodology was introduced early in the last century via the addition of hydrazine to nitriles, followed by oxidation of the resulting 1,2-dihydrotetrazine [31, 32]. However, this approach is limited to activated aromatic nitriles, and alkyl tetrazines are extremely low-yielding and sometimes impossible to obtain [27, 28]. More recently, a number of novel tetrazine synthetic schemes have been explored [33–37]. In 2012, the Devaraj group reported a Lewis acid-promoted one-pot tetrazine synthesis [34]. Nickel and zinc triflates, which likely promote amidrazone intermediate formation, have been a key additive for obtaining a series of alkyl tetrazines in moderate to high yields (Scheme 2a). This method involves the preparation of asymmetric tetrazines by tuning the ratio of two starting nitriles, and is compatible with heterocycles and most common functional groups. Thus, various tetrazine derivatives with complex structures can be obtained either directly or by further functionalization. These have included an ultrafluorogenic tetrazine **1** [38, 39], PET probes **2** [40], and thiolated tetrazine **3** utilized as an electrochemical probe [41], all of which were extremely difficult or impossible to synthesize using previous methods.

Another facet of tetrazine derivative synthesis is post-modification from simple tetrazine building blocks. The Devaraj group recently reported a new strategy for the synthesis of alkenyl tetrazine derivatives from diverse 3-mesylated-ethyl-tetrazine building blocks (Scheme 2b). Through the use of a mild elimination-Heck cascade reaction, an assorted array of π-conjugated functionalized tetrazine derivatives were obtained. Furthermore, classical fluorophore skeletons can be efficiently conjugated with tetrazines to form highly fluorogenic probes. These probes exhibit greater than 100-fold increased fluorescence after bioorthogonal reactions, rendering them an optimal alternative for applications in chemical

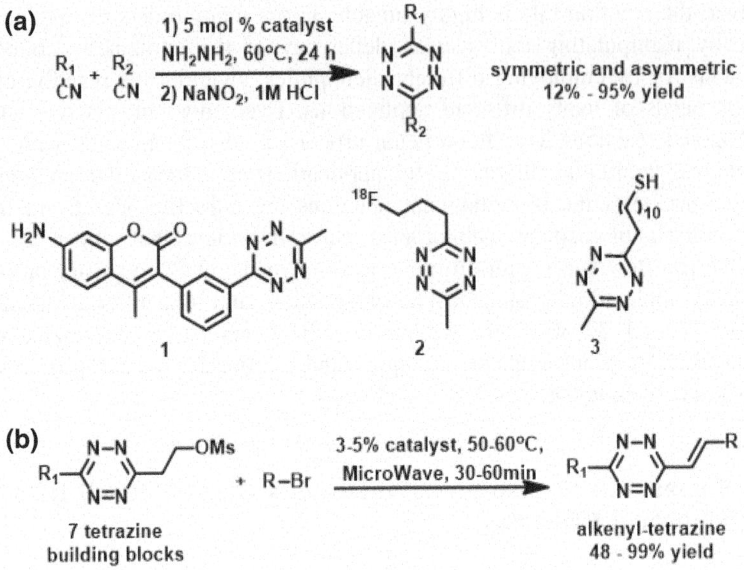

Scheme 2 **a** One-pot Lewis acid-promoted asymmetric tetrazine synthesis and corresponding functionalized tetrazine derivatives. **b** Alkenyl-1,2,4,5-tetrazine synthesis by a cascade elimination-Heck reaction

biology and biomolecular imaging [42, 43]. Similar recently reported work involved the use of Sonogashira and Stille cross-coupling on tetrazine skeletons. Here, however, the efficiency of fluorescence quenching is reduced due to the weakening of through-bond energy transfer as a result of increased donor–acceptor distance [44].

2.2 Recent Development of Strained Dienophiles for Tetrazine Bioorthogonal Chemistry

Research aimed at expanding the scope of dienophiles has generated an explosion of interest in recent years. A variety of strained alkenes or alkynes have been developed for different biomedical applications.

In 2008, Fox and co-workers reported that *trans*-cyclooctene (TCO) could be produced in gram scale from *cis*-cyclooctene by a circular photochemical isomerization procedure [45], which demonstrated an unusually fast second-order rate constant between TCO **5** and dipyridyl-tetrazine is unusually fast ($k_2 > 10^3$ M^{-1} s^{-1} in 9:1 methanol/water) [20]. As the ring strain is essential to the corresponding reaction kinetics, the authors subsequently optimized TCO by fusing a cyclopropane or dioxolane onto the cyclooctene scaffold [46, 47]. The resulting bicyclic system sterically forces the eight-membered ring into a "half-chair" conformation. These highly strained TCO compounds **6** and **7** displayed excellent rate constants of $k_2 = 3{,}300{,}000$ and 366,000 M^{-1} s^{-1}, respectively (Scheme 3a). Despite the severe strain, the dioxolane-fused TCO **7** exhibited up to

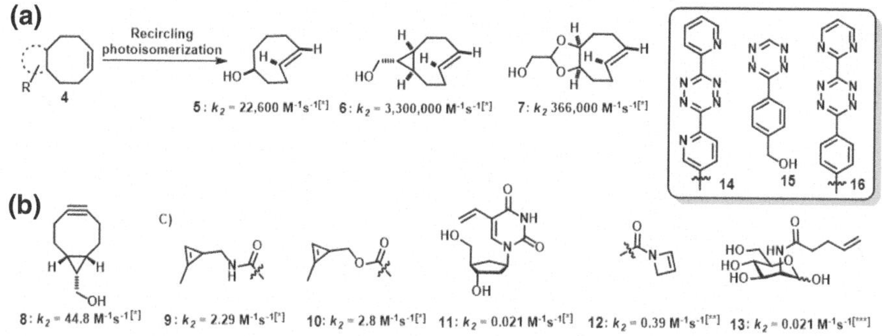

Scheme 3 Dienophiles for tetrazine bioorthogonal reaction and corresponding second-order rate constant. **a** *Trans*-cyclooctene synthesis and rate constant comparison. **b** Bicyclononyne as a dienophile. **c** Recently reported dienophile "mini-tags" and their corresponding rate constant. *k_2 was measured by reacting with tetrazine **14**; **k_2 was measured by reacting with tetrazine **15**; ***k_2 was measured by reacting with tetrazine **16**

97 % stability in human serum after 4 days [47], although compound **6** was impractical in vivo due to biologically promoted deactivation [48]. More recently however, Murrey et al. [49] demonstrated improved in vivo stability of **6** by forming complexes with $AgNO_3$. In addition, van Delft and co-workers reported the synthesis of highly strained bicyclooctynes for use in click chemistry [12]. The Wang group later utilized this dienophile in a Diels–Alder tetrazine reaction, reporting a rate constant of $k_2 = 44.8\ M^{-1}\ s^{-1}$ (Scheme 3b) [50].

In addition to rapid reaction kinetics, the size of the dienophiles is an important consideration in bioorthogonal chemistry. Various "mini-tags" have recently been developed in order to minimize structural perturbation of many biological pathways (Scheme 3c). A branch of methyl-cyclopropenes was reported by Devaraj in 2012, with favorable in vivo stability and up to 2.8 $M^{-1}\ s^{-1}$ second-order rate constants [51, 52]. Following on this work, Prescher applied the tags to carbohydrates [53, 54]. Other mini-tags, including terminal alkene-decorated biomolecules [55, 56] and azetine derivatives [57], have been reported recently for specific metabolic labeling purposes.

Other than ligations, the click-release feature is useful for applications such as drug release [58], protein decaging [59], and nucleic acid-templated signal amplification [60]. Robillard and co-workers found that the ligation intermediate between tetrazine and *trans*-cyclooctene-2-enol carbamate derivatives **17** was able to undergo a rearrangement-elimination process, resulting in a linker-cleaved product [61] (Scheme 4a). A tetrazine-mediated transfer (TMT) reaction was also reported by the Devaraj group in 2014 [43]. Specifically, 7-azabenzonorbornadiene, a strained dienophile, reacts with a tetrazine through a cascaded tetrazine ligation and retro-Diels–Alder process, resulting in two products (Scheme 4b). The release feature and mild kinetics allow TMT reactions to be employed in oligonucleotide template-driven turnover.

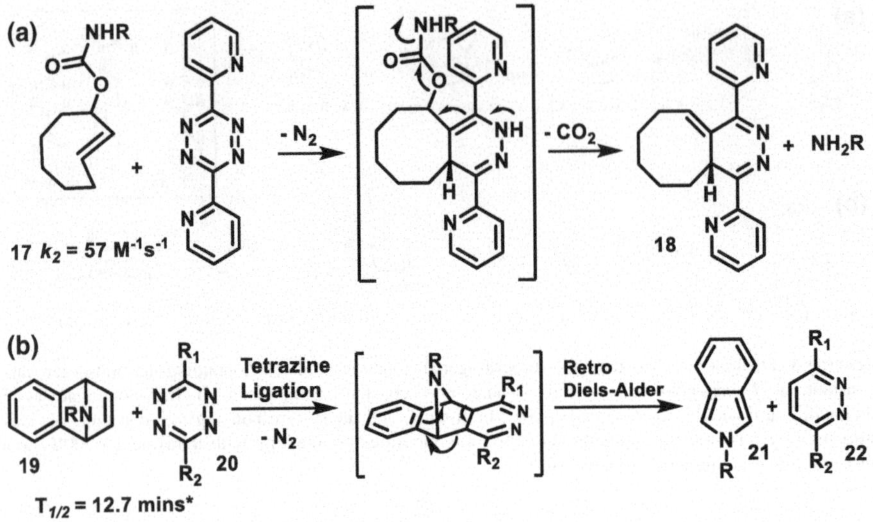

Scheme 4 **a** The rearrangement-elimination process between tetrazine and TCO-2-enol reaction. **b** The tetrazine-mediated transfer reaction. *Reaction half-life was measured in a nucleic acid-templated reaction

2.3 Recent Development of Other iEDDA Bioorthogonal Reactions

Other iEDDA reactions have also been reported within the bioorthogonal paradigm. The Lei group demonstrated the ability of the hydroxymethyl-quinoline derivative **23** to efficiently generate an active *o*-quinolinone quinone methide (*o*QQM) under physiological conditions [62]. Bioorthogonal ligations of heterodiene and vinyl thioethers can thus take place to form the product **26** (TQ ligation). The kinetic rate can be improved by increasing the electron deficiency of the quinoline precursor (Scheme 5a) [63]. In vivo stable 1,2,4-triazines have also been demonstrated as bioorthogonal handles albeit with relatively low reactivity rate (Scheme 5b) [64].

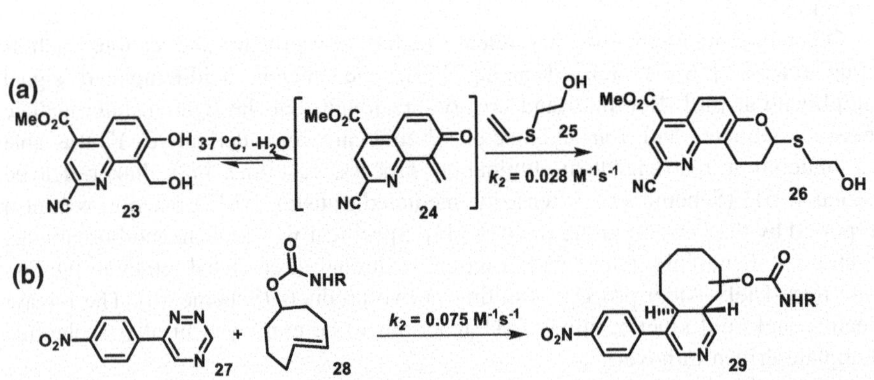

Scheme 5 **a** Reaction scheme of TQ ligation. **b** Reaction scheme of Triazine and TCO bioorthogonal reaction

3 Recent Developments of iEDDA Bioorthogonal Reactions on Biomolecules

Along with the progress in chemical research, interest has exploded recently in the adaptation of iEDDA bioorthogonal reactions to biological applications. Many kinds of biomolecules (including nucleic acid, protein, glycans, lipid, immune molecules, and bioactive small molecules) have been explored in several research veins. Numerous exciting applications have been demonstrated with success, including the pre-targeting of antibodies for imaging of tumors using PET [65, 66], protein super-resolution visualization [67], tracking of systemic live-animal glycometabolism [42], endogenous oncogenic miRNA detection [43], and bioactive molecular target identification [68]. In the following section, we will discuss the recent investigation of applications involving various biomolecules based on iEDDA bioorthogonal reactions.

3.1 Nucleic Acid Labeling and Detection by iEDDA Bioorthogonal Reaction

Nucleic acids are biological macromolecules that carry important genetic material, and play an essential role in encoding, transmitting, expressing, and regulating genetic information.

A few early examples of nucleic acid post-modification were demonstrated by the Jäschke group and others [69–72]. Dienophiles such as norbornene and TCO can be introduced into oligonucleotides during solid-phase synthesis. Site-specific labeling of DNA and RNA can then be efficiently carried out by tetrazine bioorthogonal ligations. Additionally, DNA dual labeling can be achieved simultaneously by orthogonal tetrazine ligation and click reaction (Fig. 1a). The Schorr group recently created synthetic RNA oligonucleotides with a norbornene tag, which can be labeled by a tetrazine fluorogenic probe in mammalian cells using the same strategy (Fig. 1b) [73].

The utilization of template-driven reactions with fluorogenic probes is a powerful technique for nucleic acid detection [74–77]. The first use of tetrazine bioorthogonal chemistry to accomplish this aim was reported by the Devaraj group [78]. A pair of tetrazine and methyl-cyclopropene probes are brought into spatial proximity in the presence of an antisense oligo template. The use of the template increases the

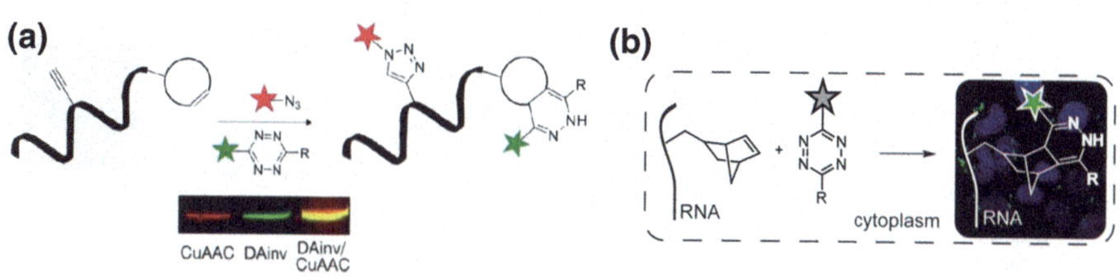

Fig. 1 **a** DNA post-modification by tetrazine ligation and click reaction. **b** Live cell synthetic RNA target image using tetrazine fluorogenic probe. **a** Adapted with permission from [70]. Copyright (2012) American Chemical Society. **b** Adapted with permission from [73]. Copyright (2014) American Chemical Society

effective molarity, and a rapid bioorthogonal reaction takes place, resulting in a large increase in fluorescence (Fig. 2a). However, due to the nature of tetrazine–cyclopropene reactions, the ligation product has a higher affinity for the oligo template than the corresponding precursors. This feature hampers further reaction turnover and signal amplification, which is essential for endogenous low-concentration nucleic target detection [60, 75, 79–83]. The Devaraj group further optimized their probe using a norbornadiene derivative and alkenyl-tetrazine fluorogenic probe. The tetrazine-mediated transfer reaction facilitated reaction turnover and fluorogenic signal amplification by virtue of the cascade retro-Diels–Alder unlinking of the two oligo probes after the initial tetrazine ligation (Fig. 2b). With this strategy, endogenous oncogenic miRNA targets can be detected in both cell lysates and live cells. Remarkably, this method is highly sensitive for similar target discrimination. Single base-pair mismatched target miRNA can be distinguished with up to 24-fold less signal read out [43].

The utilization of metabolic probes is a key strategy for the decryption of intracellular life processes. Luedtke and co-workers developed a vinyl-decorated deoxyuridine **11**(VdU) that was able to be incorporated by endogenous enzymes into DNA synthesis and visualized by fluorescent tetrazine conjugates [55]. Additionally, VdU displayed lowered cytotoxicity and caused less DNA damage and cell cycle arrest compared to other "clickable" metabolic nucleic probes such as 5-ethynyl-2′-deoxyuridine (Fig. 3).

3.2 Site-Specific Protein/Peptide Labeling by iEDDA Bioorthogonal Reaction

Site-specific incorporation of a bioorthogonal reporter into functional proteins has attracted a great deal of attention over the last two decades [25, 84]. It is an ideal strategy for understanding protein function and dynamics, probing natural systems, uncovering disease mechanisms, and revealing important structure–activity relationships. Several techniques have been utilized, including fluorescent protein

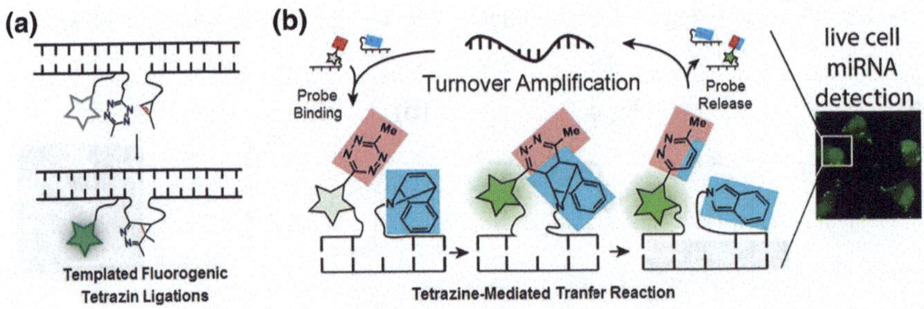

Fig. 2 **a** Schematic of templated fluorogenic tetrazine ligations. **b** Schematic of template tetrazine-mediated transfer reaction for live cell miRNA detection. **a** Adapted from Ref. [78] with permission from Oxford University Press; **b** Adapted with permission from [43]. Copyright (2014) American Chemical Society

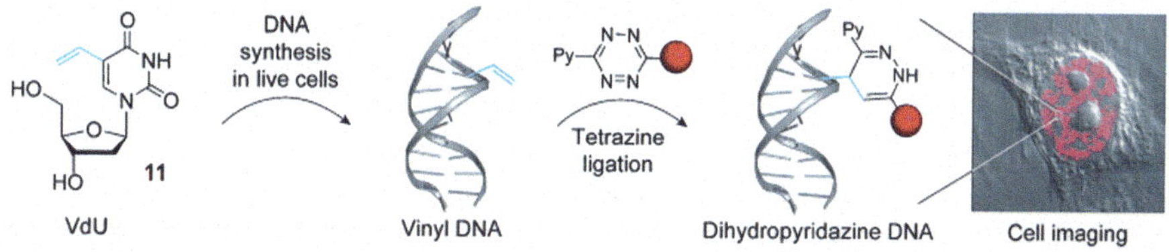

Fig. 3 Cellular DNA imaging using metabolic nucleic acid probe. Figure reprinted from Ref. [55] with permission from John Wiley and Sons

fusion [85–87], incorporation of reporters by ligases [88–90], site-specific self-labeling reporters [91, 92], and unnatural amino acid incorporation [93–95].

3.2.1 *Introducing Bioorthogonal Reporters into Proteins Via Genetic Code Expansion*

Among the array of bioorthogonal reporters discussed, genetic unnatural amino acid incorporation strategies have had one of the most dramatic impacts upon bioorthogonal chemistry [84]. An unnatural amino acid, bearing a bioorthogonal reaction handle, can be recognized by a tRNA-synthetase to form the tRNA-unnatural amino acid complex. Thus, the unnatural amino acid can be encoded into peptide synthesis, resulting in a modified protein with a bioorthogonal reporter at the desired site (Scheme 6a). A panel of unnatural amino acids has been developed in recent years for this strategy (Scheme 6b).

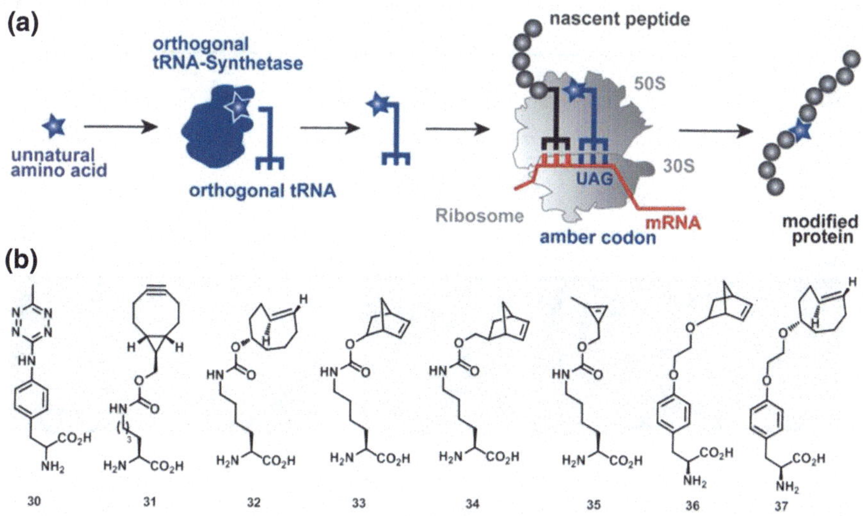

Scheme 6 **a** Schematic of strategy for expanding the genetic code. **b** Unnatural amino acid bearing a tetrazine reaction group for site-specific incorporation via genetic code expansion. **a** Adapted with permission from [84], Copyright (2014) American Chemical Society

The Mehl group demonstrated that the tetrazine modified phenylalanine **30** was recognized by the *Methanococcus jannaschii* (*Mj*) tyrosyl-tRNA synthetase (RS)/ $tRNA_{CUA}$ pair, resulting in tetrazine-incorporated proteins [96]. Interestingly, when the tetrazine group is spatially close to green fluorescent protein (GFP, ~12 Å), the chromophore is slightly quenched, and an 11-fold fluorescent turn-on can be observed after reacting with a TCO partner. The Chin group elegantly developed a variety of tRNA synthetase and modified tRNA pairs capable of incorporating diverse dienophiles (bicyclooctyne **31**, *trans*-cyclooctene **32**, norbornene **33** and **34**, and cyclopropene **35**) into protein expression in both *E. coli* and mammalian cells. Furthermore, rapid site-specific protein labeling was accomplished through reaction with fluorogenic tetrazine [97, 98]. The Lemke group and others reported successful one-pot double-labeling of protein by the introduction of two orthogonal functional groups on different amino acids [99–101]. Shortly thereafter, the Chin group demonstrated the evolution of novel quadruplet-decoding transfer RNAs, which greatly improved the efficiency of unnatural amino acid incorporation [102].

Super-resolution cellular ultrastructure imaging and specific tissue labeling based on the genetic code expansion strategy have been recently reported [67, 103]. Functionalized proteins with different cellular localizations can be efficiently labeled by tetrazine ligations using photostable fluorophores suitable for stochastic

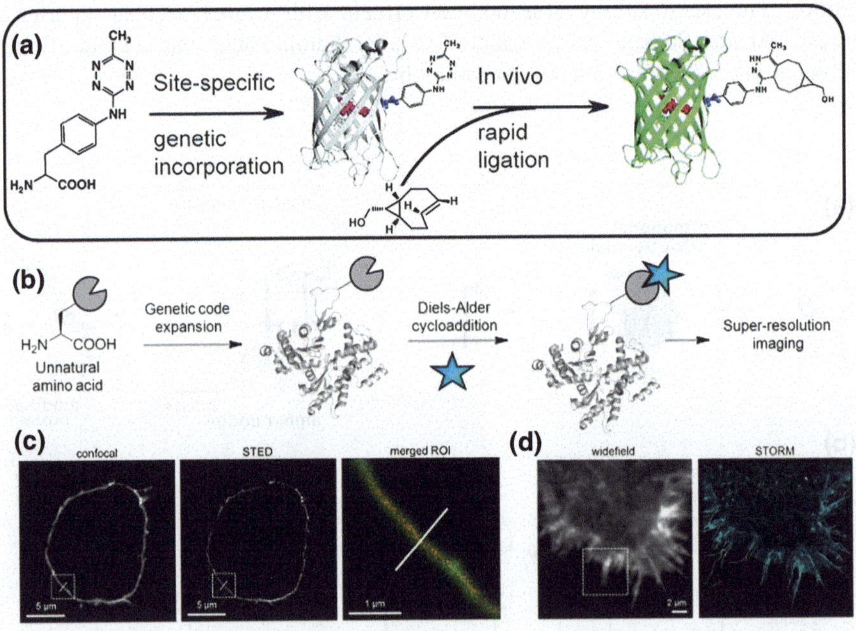

Fig. 4 **a** Genetic site-specific incorporation with unnatural tetrazine-phenylalanine resulting in turn-on of green fluorescent protein (GFP). **b** Schematic of genetic code strategy for super-resolution cellular protein imaging. **c** Confocal and stochastic optical reconstruction microscopy (STORM) imaging of cell-surface EGFR in HEK293T, with zoom-in imaging of region of interest (ROI). **d** Wide-field and corresponding STORM image of intracellular actin. **a** Adapted with permission from [96], Copyright (2012) American Chemical Society. **b** Adapted with permission from [67], Copyright (2015) American Chemical Society

optical reconstruction microscopy (STORM) imaging. These studies revealed the spatial details of surface/interior proteins. Furthermore, actin with distinct morphologies at different cellular locations can be observed with nanometer-scale refined features (Fig. 4d).

3.2.2 Introduction of Bioorthogonal Reporters Via Other Strategies

Other protein labeling methods developed within the last few years have taken advantage of the inverse electron-demand Diels–Alder reaction of tetrazines. Ting and co-workers utilized their PRIME (probe incorporation mediated by enzymes) strategy to decorate proteins with TCO [104]. A lipoic acid ligase (LplA) with a single mutation at Trp37 can efficiently ligate TCO **38** onto target proteins, which can be visualized by a second-step labeling reaction with tetrazine fluorogenic probes (Fig. 5a). This robust method enables the investigation of a variety of proteins labeled at various cellular locations, including surface lipoproteins, neuroligin, actin, and vimentin (Fig. 5b–d). A direct peptide/protein labeling protocol was recently demonstrated by the Smith group [105], in which *S*-dichlorotetrazine was able to be introduced between two proximate cysteine sulfhydryl groups by nucleophilic displacement. The resulting tetrazine peptide/protein can then be labeled by a complementary dienophile (Fig. 6). Initial peptide/proteins can be regenerated by a UV-catalyzed tetrazine decomposition release process. However, proteins lose activity during the cleavage event, implying that the chemistry is no longer truly bioorthogonal.

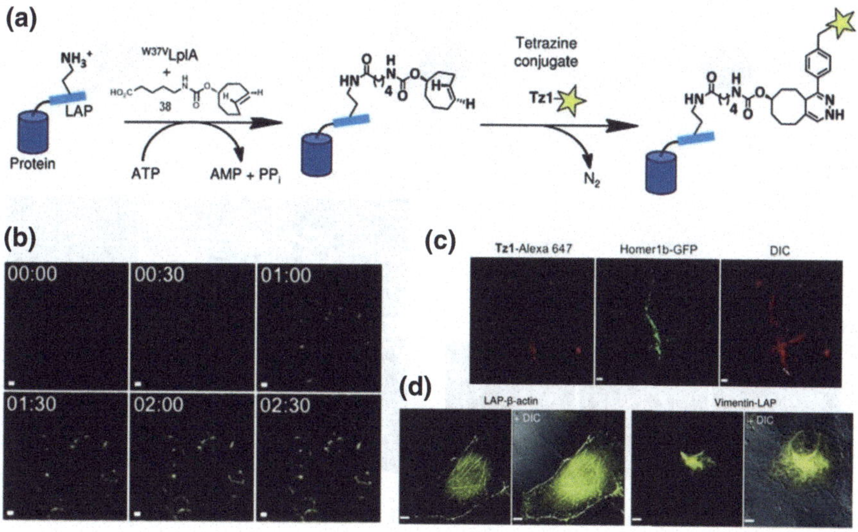

Fig. 5 **a** Schematic of PRIME (probe incorporation mediated by enzymes) strategy for protein labeling. **b** Real-time imaging of cell surface lipoprotein. **c** GFP-fused neuroligin were imaged by tetrazine-Alexa Fluor 647. **d** Intracellular actin and vimentin imaging by reaction with Tz-fluorescein diacetate. Figures adapted with permission from [104], Copyright (2015) American Chemical Society

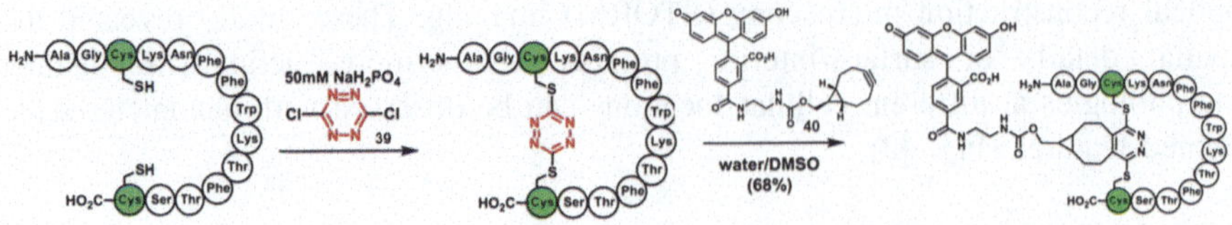

Fig. 6 Peptide/protein labeling by direct *S*-dichlorotetrazine insertion and subsequent iEDDA reaction. Figures adapted with permission from [105], Copyright (2015) American Chemical Society

3.3 Functional Protein Labeling Via Post-modification

In addition to the bottom-up genetic code expansion technique, functional proteins (typically antibodies) can undergo post-modification with a bioorthogonal reporter for subsequent labeling/imaging. Antibodies are an attractive tool given their highly specific binding to the corresponding target antigen, and thus can be used as a highly specific bioorthogonal reporter. Disease-related antigen overexpression can also be detected or imaged utilizing bioorthogonal reactions (Fig. 7a).

Tetrazine bioorthogonal small-molecule fluorescent probes have been widely developed for labeling pre-targeted antibodies. Pioneer work was reported by the Weissleder group in 2009. A549 cancer cells that over-express epidermal growth factor receptor (EGFR) were pre-targeted by cetuximab-TCO conjugates. The cells

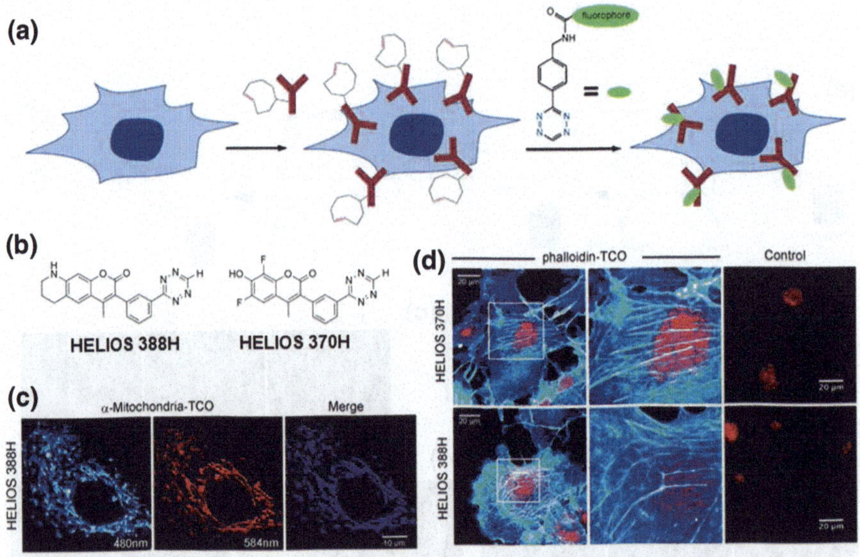

Fig. 7 **a** Schematic of pre-targeted antibody strategy. **b** Chemical structure of fluorogenic coumarin–tetrazine probes. **c** RFP-tagged mitochondria were imaged by HELIOS 388H probe. **d** Actin imaging: Phalloidin-TCO targeted actin was imaged by HELIOS 388H probe. Control experiments were carried out at same HELIO probe concentration without phalloidin-TCO. **a** Reprinted from Ref. [106], with permission from John Wiley and Sons. **c**, **d** Reprinted from Ref. [39], with permission from John Wiley and Sons

were then labeled by a serum-stable tetrazine bearing a far-red fluorophore, VT680, for visualization [106]. Using a similar strategy, Hilderbrand demonstrated orthogonality between tetrazine ligations and SPAAC for labeling different antibody targets in parallel [107]. One promising feature of tetrazine ligations is their fluorogenic nature, which significantly improves image resolution by increasing signal over background. A series of highly fluorogenic tetrazine probes have been developed with hundreds- to 11,000-fold turn-on after undergoing a tetrazine ligation [37–39, 108]. These probes, such as coumarin–tetrazine HELIOS probes (structure shown in Fig. 7b), are capable of labeling cell surface antibodies without a wash step, thus enabling elucidation of the fine structure and locale of intracellular targets (Fig. 7c, d).

In addition to small-molecule fluorescent probes, diagnostic nanoparticles have been reported for protein labeling, as described by Bawendi and co-workers [109]. Cellular labeling of EGF-tetrazine was efficiently achieved by bioorthogonal reaction with norbornene-quantum dots using one of two strategies. Cells are either directly exposed to EGF-QDs conjugates (one-step strategy: BOND-1, Fig. 8a) or

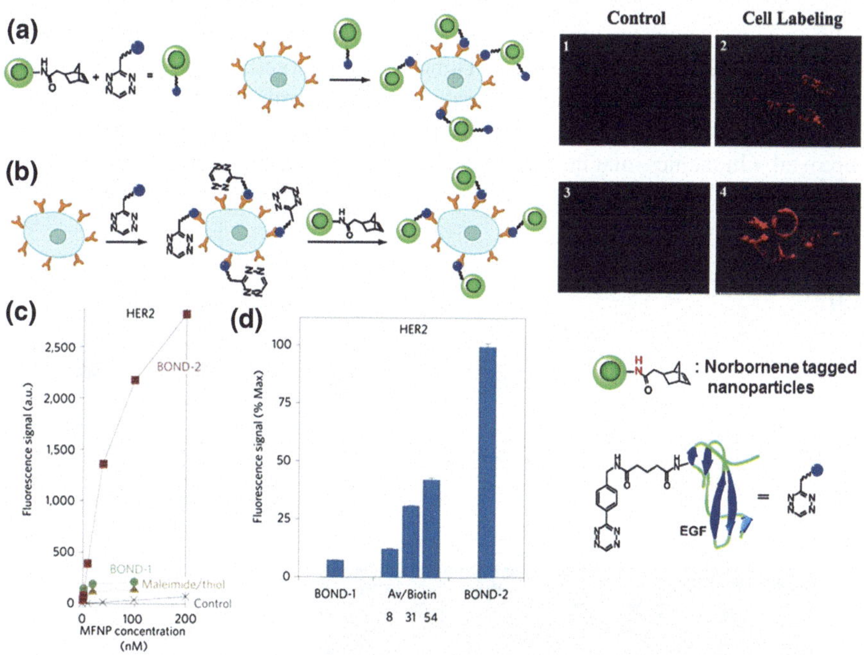

Fig. 8 **a** Schematic of BOND-1 strategy, and corresponding live cell EGFR labeling. **b** Schematic of BOND-2 strategy, and corresponding live cell EGFR labeling. Control experiments were carried out with the same QDs but no norbornene modification. **c, d** Comparison of different nanoparticle labeling strategies: **c** HBR-3 cells were labeled with different concentrations of MFNP using the two-steps BOND-2 or direct MFNP immunoconjugates, and the fluorescence signal was measured using flow cytometry. Control samples were incubated with Tz–MFNP only. **d** Nanoparticle binding increased with biotin loading but remained lower than BOND-2. Figure adapted from Refs. [109, 110] with permission from the (2010) American Chemical Society and Nature Publishing Group

are first incubated with tetrazine-decorated EGF, followed by the addition of norbornene-QDs (two-step strategy: BOND-2, Fig. 8b). Independently, the Weissleder group systematically compared the efficiency of two tetrazine labeling strategies (BOND-1 and BOND-2) with other classical methods including avidin/biotin and maleimide/thiol chemistry [110]. Using highly sensitive magneto-fluorescent nanoparticles as a readout, the authors demonstrated that BOND-1 was comparable to other covalent conjugation methods (Fig. 8c), while BOND-2 significantly improved labeling efficiency by site-specific amplification of nano-material binding (Fig. 8d). They hypothesized that BOND-2 is able to decorate a higher valency of nanoparticles by tetrazine-TCO ligation, while immunoconjugates may be restrained by steric hindrance. Moreover, BOND-1 is able to obtain less than one nanoparticle per antibody due to potential cross-linking and aggregation.

The use of the BOND-2 platform can significantly enhance the detection sensitivity and efficiency of nanomaterials, and many exciting biomedical applications have been launched in this area. For example, the cytoplasmic and nuclear protein target expression levels can be quantitatively distinguished by magnetic and fluorescent tetrazine-tagged nanoparticles [111], and rare cell enumeration can be immunomagnetically tagged and detected by micro-magnetic resonance sensors [112] or imaged by quantum dot immunoconstructs [113]. Valliant and co-workers recently reported immune-ultrasonic molecular imaging using the tetrazine BOND-2 strategy. Tumor cell pre-targeting with a TCO antibody can be visualized by tetrazine-decorated gas-filled microbubbles (MB_{Tz}) as a contrast agent (Fig. 9) [114]. Both in vitro and mice experiments have shown improved ultrasonic imaging through the use of a pre-targeting bioorthogonal strategy; however, the stability and efficiency of MB_{Tz} must be optimized for future clinical studies.

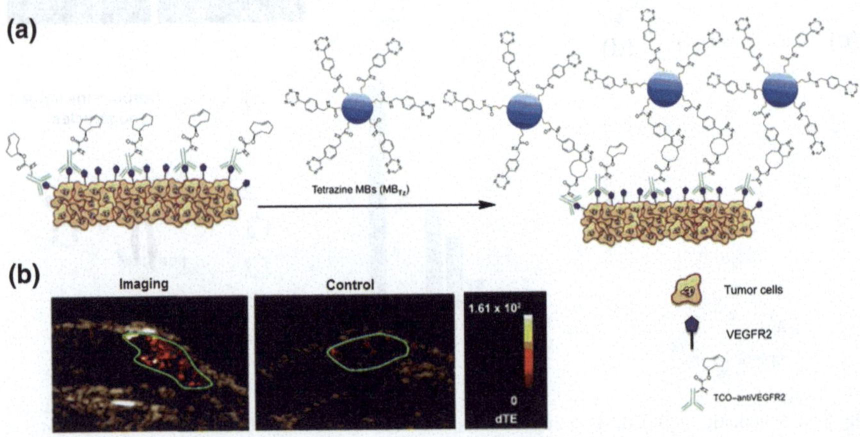

Fig. 9 **a** Localization of microbubbles to tumor cells for ultrasound imaging through pre-targeting and bioorthogonal chemistry. **b** Transverse color-coded parametric nonlinear contrast-mode ultrasound images acquired 4 min after intravenous administration of MB_{Tz} to SKOV-3 human adenocarcinoma murine tumor model [VEGFR2(+)] pre-targeted with TCO-antiVEGFR2; control experiments were carried out in the same model without the antibody. Figures reprinted from Ref. [114], with permission from John Wiley and Sons

There have also been many inspiring advancements in the area of in vivo radionuclide functional imaging. In recent years, several molecular imaging probes for single-photon emission computed tomography (SPECT) or positron emission tomography (PET) were efficiently obtained by tetrazine ligation [40, 65, 115–117]. More importantly, dienophile-tagged antibodies can accumulate specifically at the targeted region. After the excess antibodies clear, a tetrazine isotope chaser can react with the pre-targeted antibody. Benefited by the rapid kinetics of tetrazine bioorthogonal labeling, short-lived isotope probes can efficiently image antibody distribution while excess probe is cleared rapidly by the blood. Robillard and co-workers reported the first SPECT imaging using a [^{111}In]-tetrazine probe to visualize TAG72 antigen overexpression in solid tumors [118]. As short-lived ^{18}F probes are commonly used in clinical diagnosis, Devaraj et al. systematically investigated the effects of the bioorthogonal reaction rate and isotope probe clearance rate for in vivo pre-targeted PET imaging. After careful computational modeling and confirmatory experimental investigation, they demonstrated that the pharmacokinetics of the isotope probes must remain at a sufficient concentration for in vivo bioorthogonal reactions (Fig. 10a). To this end, they developed an oligomeric dextran tetrazine ^{18}F probe for pre-targeted PET imaging to regulate clearance rates of their probe. Colon cancer xenografts were targeted by TCO-modified anti-A33 monoclonal antibodies. After clearance, significant uptake of the injected ^{18}F-tetrazine probe was observed at the tumor site, as visualized on PET imaging. The imaging signal at the tumor site was observed to a much lesser extent when the antibodies did not contain the *trans*-cyclooctene or when targeting tumors that lacked the corresponding antigen (Fig. 10b, c) [66]. Inspired by the works just described, other studies have reported similar accomplishments using other longer-half-life isotopes [119–121].

3.4 iEDDA Bioorthogonal Reactions on Lipids

Lipid imaging and labeling using bioorthogonal chemistry has also attracted widespread interest in recent years. Several exciting applications have been

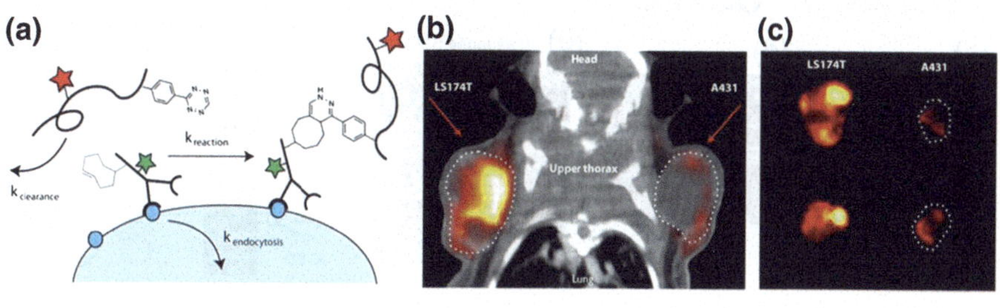

Fig. 10 **a** Balance of different kinetic parameters for in vivo bioorthogonal reaction. The secondary tetrazine probe reacts with TCO antibodies at a given rate ($k_{reaction}$), which competes with the clearance of the tetrazine probe from the body ($k_{clearance}$) and internalization of the antibody ($k_{endocytosis}$). **b** PET/CT fusion images of mouse with A431 and LS174T tumors pre-targeted with anti-A33 TCO monoclonal antibodies followed by ^{18}F-tetrazine probe. *Arrows* indicate location of tumors. **c** Autoradiography of representative 1-mm LS174T or A431 tumor slices after multi-step targeting. Figure reproduced with permission from Ref. [66]

explored, including the monitoring of lipid transportation, metabolic labeling, and high-throughput analysis of protein lipidation [122]. Some of the first conceptual work in this area using tetrazine bioorthogonal chemistry was reported by the Devaraj group in 2012 [51]. Phospholipid distribution in live cells can be visualized =using cyclopropene-modified phospholipids and tetrazine fluorogenic probes (Fig. 11a).

The Schepartz group very recently reported their lipid-based strategy for Golgi super-resolution imaging [123]. *Trans*-cyclooctene-decorated ceramide lipids (*Cer-TCO*) were selectively localized at Golgi [124] and visualized by a highly reactive tetrazine near-IR probe **SiR-Tz** (Fig. 11b). Systematic organelle and cell compatibility assays were performed, demonstrating that neither the probe nor ligation product disturbed traffic through or within the Golgi. The ligation reaction is also nontoxic to cells for hours. More importantly, compared to commercial fluorescent lipids, the resulting "vital dye" is highly photostable, enabling prolonged super-resolution stimulated emission depletion (STED) imaging in live cells (Fig. 11c–e).

3.5 iEDDA Bioorthogonal Reaction on Glycans

Glycans play myriad roles in cellular physiology, including cell signal translation, proliferation, differentiation, and migration. In addition, glycosylation disorder is associated with a number of diseases [125]. As such, there has been tremendous interest within recent years in the visualization of glycans in live cells or animals to understand glycobiology [6, 126, 127].

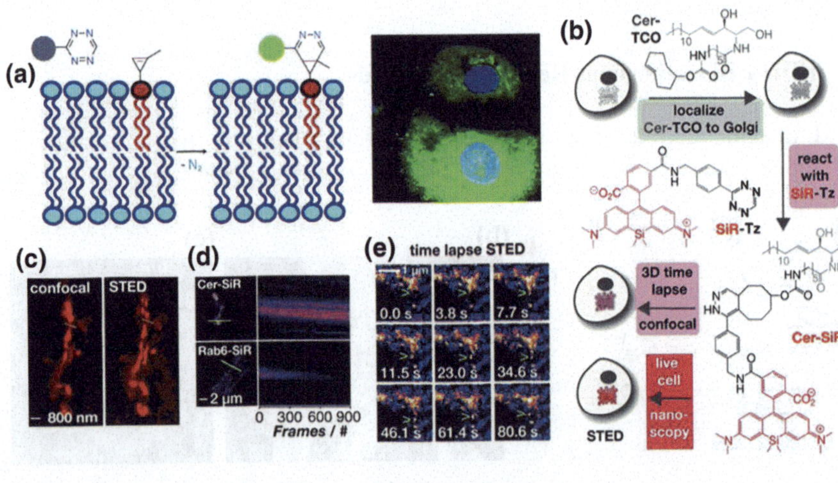

Fig. 11 **a** Cyclopropene-tagged phospholipid for live cell lipid image. **b** Schematic description of high-density labeling of the Golgi using a two-step procedure. **c** Comparison of confocal and stimulated emission depletion (STED) images of the live cell Golgi. **d** Comparison of kymographs of fixed-cell STED image labeled by Cer-SiR and or Rab6-SNAP-SiR, respectively; note that the Cer-SiR is much more photostable. **e** Time-lapse super-resolution STED depletion of vesicle budding and trafficking out of the Golgi (*green arrowhead*). Figures reprinted from Refs. [51, 123] with permission from John Wiley and Sons

Prescher and co-workers showed that cyclopropenes conjugated to sialic acid could be metabolized by cells and incorporated into cell surface glycans [53]. The resulting glycans with the incorporated mini-tags can thus be labeled by Diels–Alder reactions with tetrazine-fluorophore probes. Devaraj and co-workers reported the successful incorporation of cyclopropene-mannosamine derivatives into human cancer cell biosynthetic pathways as a metabolic probe (Fig. 12a) [128]. The use of terminal alkene-tagged mannosamine as a metabolic probe for cell surface glycan visualization was reported by the Wittmann group (Fig. 12b) [56]. All of these works demonstrated that metabolic alkene and azide sugars could be simultaneously labeled by complementary bioorthogonal groups, thereby encouraging future multi-target imaging studies.

The first example of systemic live-animal imaging using a bioorthogonal fluorogenic probe was reported by Bertozzi and co-workers [42]. Cyclooctyne-tagged sialic acid (BCNSia), which has high reactivity with tetrazine and in vivo stability, can also be incorporated into the glycan biosynthesis pathway as a metabolic probe. Zebrafish embryos were injected with membrane-impermeable BCNSia at the 1–8 cell stage. BCNSia was well-diffused, along with vein plexus formation. Subsequently, mixtures of fluorogenic tetrazine probe **41** and a far-red nonspecific tracer were injected into zebrafish vasculature, resulting in an elaborate systemic sialylated imaging of the resulting glycoconjugates. This powerful method enables the visualization of sugar metabolism levels at different regions. Moreover, the image details at the sub-cellular level uncovered novel sialylated structures in developing zebrafish, which should guide future research in this area (Fig. 13).

3.6 iEDDA Bioorthogonal Reaction on Bioactive Small Molecules

Visualizing the dynamic distribution of bioactive small molecules and identifying the corresponding targets are crucial in biomedical studies and drug discovery. Bioactive molecules modified with a bioorthogonal reporter can be labeled by a

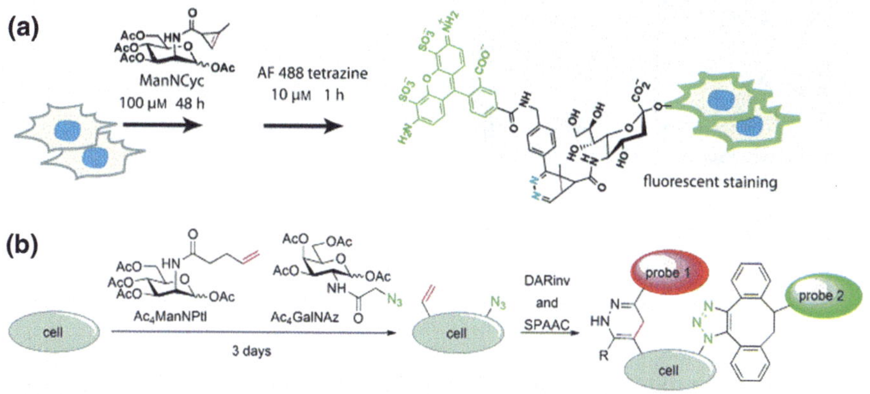

Fig. 12 **a** Cyclopropene-tagged mannosamine for glycan metabolic bioorthogonal imaging. **b** Azide- and terminal alkene-decorated mannosamine for glycan metabolic bioorthogonal labeling. Figures reprinted from Refs. [56, 128] with permission from John Wiley and Sons

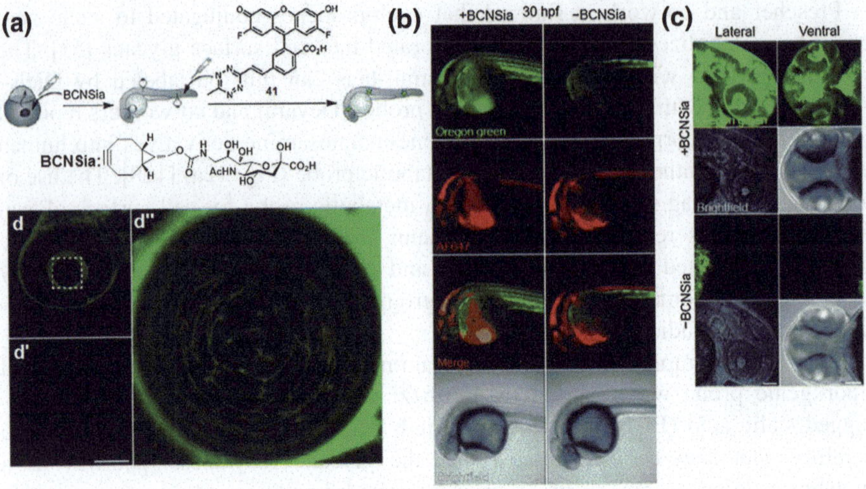

Fig. 13 **a** Scheme showing injections of BCNSia and **41**; embryos were injected with BCNSia or vehicle at the 1–8 cell stage and then injected with **41** in the caudal vein 1 h prior to imaging. **b** Projection images of 30 embryos treated with BCNSia or vehicle; **41** was co-injected with AF647-NH_2 to map the vasculature. *Scale bar*= 200 mm. **c** Projection images (×20) of 48-h-post-fertilization embryos from lateral and ventral views. **d**, **d′** Lateral view of the eye and lens (**d″**). Zoom-in of the *dashed box* in (**d**). Figures reprinted from Ref. [42], with permission from John Wiley and Sons

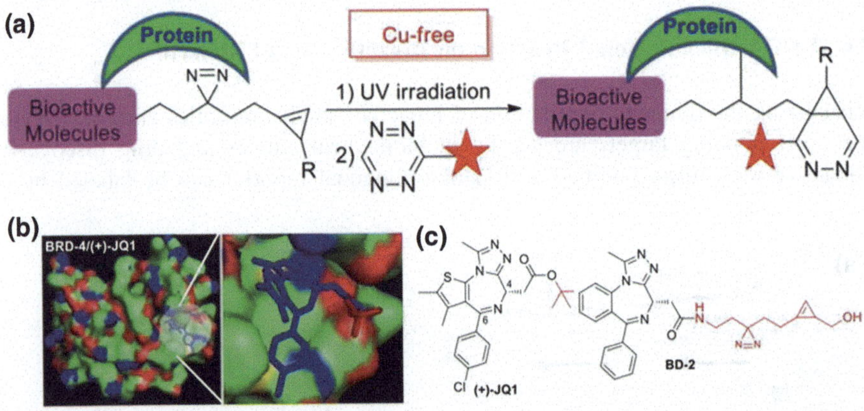

Fig. 14 **a** Schematic of drug-profiling photoaffinity labeling approaches using cyclopropenes as chemically tractable tags suitable for tetrazine bioorthogonal chemistry. **b** The X-ray complex (PDB code: 3MXF) of BRD-4 protein with known bioactive compound (+)-JQ1 showing the active site binding (*tert*-butyl group is marked in *red*). **c** Chemical structures of (+)-JQ1 and **BD-2** probes for photoaffinity labeling. Figures adapted with permission from [68], Copyright (2014) American Chemical Society

complementary probe reporting on locale and concentration of the bioactive molecule of interest. More importantly, the decoration does not change the affinity behavior of the biomolecule. The first successful use of a tetrazine probe was reported by Weissleder and co-workers [108]. Taxol, a well-studied anti-cancer agent known for its ability to stabilize microtubules in cells, was chosen as a model compound. TCO-modified taxol exhibiting bioactivity similar to its parent compound was introduced into cells by routine incubation. The intracellular spatial distribution of taxol was imaged by tetrazine fluorogenic probes. Studies with kinase dynamic imaging and quantification were also reported using a similar strategy [129, 130].

Recently a group led by Yao developed a minimalist linker technique enabling simultaneous imaging of the bioactive molecule and covalent cross-linking to the adjacent target (Fig. 14) [68]. The minimalist linker contains a cyclopropene for tetrazine ligation and a diazirine group for UV cross-linking purposes. The affinity of the molecule with the target can be detected by tetrazine fluorogenic probes both in vitro and in live cells. In addition, UV covalent ligation allows for affinity pull-down assays of the target for further proteome profiling. The probe's dual capability will undoubtedly facilitate future drug target studies [131].

4 Conclusion

Tetrazine reactions, the most widely used representative of the inverse electron-demand Diels–Alder bioorthogonal family, has emerged as a powerful tool for decryption of biological messages from organisms. As illustrated in this review, the scope of reaction precursors has been greatly expanded, with tunable kinetics, versatile functions, and improved stability. Reactions have been efficiently carried out on most classes of biomolecules for various important investigations. A wide range of detection methods and other bioorthogonal chemistries can be combined into tetrazine bioorthogonal chemistry to meet diverse biomedical and clinical research needs. While much exciting work has been accomplished to date, there are still many aspects of this technology that must be optimized in the future. Of these, the kinetics of mini-tagged dienophiles must be improved, fluorogenic near-infrared tetrazine probes for deep tissue imaging are yet undiscovered, and novel multifunctional probe development is an attractive area for new applications. Further investigation within the field of inverse electron-demand Diels–Alder bioorthogonal chemistry should provide multiple prospects for fundamental biomedical research and a vast array of clinical investigative and diagnostic applications.

Acknowledgments The authors gratefully acknowledge S. Alexander and C.Y. Zhou for their many helpful discussions and suggestions. We thank the University of California, San Diego, for financial support.

References

1. Prescher JA, Bertozzi CR (2005) Nat Chem Biol 1:13–21
2. Sletten EM, Bertozzi CR (2009) Angew Chem Int Ed 48:6974–6998
3. Boyce M, Bertozzi CR (2011) Nat Methods 8:638–642
4. Sletten EM, Bertozzi CR (2011) Acc Chem Res 44:666–676
5. Agard NJ, Baskin JM, Prescher JA, Lo A, Bertozzi CR (2006) ACS Chem Biol 1:644–648

6. Baskin JM, Prescher JA, Laughlin ST, Agard NJ, Chang PV, Miller IA, Lo A, Codelli JA, Bertozzi CR (2007) Proc Natl Acad Sci 104:16793–16797
7. Codelli JA, Baskin JM, Agard NJ, Bertozzi CR (2008) J Am Chem Soc 130:11486–11493
8. Jewett JC, Bertozzi CR (2010) Chem Soc Rev 39:1272–1279
9. Jewett JC, Sletten EM, Bertozzi CR (2010) J Am Chem Soc 132:3688–3690
10. Saxon E, Armstrong JI, Bertozzi CR (2000) Org Lett 2:2141–2143
11. Saxon E, Bertozzi CR (2000) Science 287:2007–2010
12. Dommerholt J, Schmidt S, Temming R, Hendriks LJA, Rutjes FPJT, van Hest JCM, Lefeber DJ, Friedl P, van Delft FL (2010) Angew Chem Int Ed 49:9422–9425
13. Ning X, Temming RP, Dommerholt J, Guo J, Ania DB, Debets MF, Wolfert MA, Boons G-J, van Delft FL (2010) Angew Chem Int Ed 49:3065–3068
14. Sanders BC, Friscourt F, Ledin PA, Mbua NE, Arumugam S, Guo J, Boltje TJ, Popik VV, Boons G-J (2011) J Am Chem Soc 133:949–957
15. McGrath NA, Raines RT (2012) Chem Sci 3:3237–3240
16. Lim RKV, Lin Q (2011) Acc Chem Res 44:828–839
17. Yu Z, Pan Y, Wang Z, Wang J, Lin Q (2012) Angew Chem Int Ed 51:10600–10604
18. Ramil CP, Lin Q (2013) Chem Commun 49:11007–11022
19. Yu Z, Ohulchanskyy TY, An P, Prasad PN, Lin Q (2013) J Am Chem Soc 135:16766–16769
20. Blackman ML, Royzen M, Fox JM (2008) J Am Chem Soc 130:13518–13519
21. Devaraj NK, Weissleder R, Hilderbrand SA (2008) Bioconj Chem 19:2297–2299
22. Devaraj NK, Weissleder R (2011) Acc Chem Res 44:816–827
23. Šečkutė J, Devaraj NK (2013) Curr Opin Chem Biol 17:761–767
24. Debets MF, van Hest JCM, Rutjes FPJT (2013) Org Biomol Chem 11:6439–6455
25. Jing C, Cornish VW (2011) Acc Chem Res 44:784–792
26. Borrmann A, van Hest JCM (2014) Chem Sci 5:2123–2134
27. Saracoglu N (2007) Tetrahedron 63:4199–4236
28. Clavier G, Audebert P (2010) Chem Rev 110:3299–3314
29. Selvaraj R, Fox JM (2013) Curr Opin Chem Biol 17:753–760
30. Knall A-C, Slugovc C (2013) Chem Soc Rev 42:5131–5142
31. Hofmann KA, Ehrhart O (1912) Ber Dtsch Chem Ges 45:2731–2740
32. Curtius T, Hess A (1930) Journal für Praktische Chemie 125:40–53
33. Karver MR, Weissleder R, Hilderbrand SA (2011) Bioconj Chem 22:2263–2270
34. Yang J, Karver MR, Li W, Sahu S, Devaraj NK (2012) Angew Chem Int Ed Engl 51:5222–5225
35. Liu H, Wei Y (2013) Tetrahedron Lett 54:4645–4648
36. Wang D, Chen W, Zheng Y, Dai C, Wang L, Wang B (2013) Heterocycl Commun 19:171
37. Wu H, Yang J, Seckute J, Devaraj NK (2014) Angew Chem 53:5805–5809
38. Carlson JC, Meimetis LG, Hilderbrand SA, Weissleder R (2013) Angew Chem Int Ed 52:6917–6920
39. Meimetis LG, Carlson JCT, Giedt RJ, Kohler RH, Weissleder R (2014) Angew Chem Int Ed 53:7531–7534
40. Denk C, Svatunek D, Filip T, Wanek T, Lumpi D, Fröhlich J, Kuntner C, Mikula H (2014) Angew Chem Int Ed 53:9655–9659
41. Ehret F, Wu H, Alexander SC, Devaraj NK (2015) J Am Chem Soc 137:8876–8879
42. Agarwal P, Beahm BJ, Shieh P, Bertozzi CR (2015) Angew Chem Int Ed 54:11504–11510
43. Wu H, Cisneros BT, Cole CM, Devaraj NK (2014) J Am Chem Soc 136:17942–17945
44. Wieczorek A, Buckup T, Wombacher R (2014) Org Biomol Chem 12:4177–4185
45. Royzen M, Yap GPA, Fox JM (2008) J Am Chem Soc 130:3760–3761
46. Taylor MT, Blackman ML, Dmitrenko O, Fox JM (2011) J Am Chem Soc 133:9646–9649
47. Darko A, Wallace S, Dmitrenko O, Machovina MM, Mehl RA, Chin JW, Fox JM (2014) Chem Sci 5:3770–3776
48. Rossin R, van den Bosch SM, ten Hoeve W, Carvelli M, Versteegen RM, Lub J, Robillard MS (2013) Bioconj Chem 24:1210–1217
49. Murrey HE, Judkins JC, Am Ende CW, Ballard TE, Fang Y, Riccardi K, Di L, Guilmette ER, Schwartz JW, Fox JM, Johnson DS (2015) J Am Chem Soc 137:11461–11475
50. Chen W, Wang D, Dai C, Hamelberg D, Wang B (2012) Chem Commun 48:1736–1738
51. Yang J, Seckute J, Cole CM, Devaraj NK (2012) Angew Chem Int Ed Engl 51:7476–7479
52. Yang J, Liang Y, Šečkutė J, Houk KN, Devaraj NK (2014) Chem A Eur J 20:3365–3375

53. Patterson DM, Nazarova LA, Xie B, Kamber DN, Prescher JA (2012) J Am Chem Soc 134:18638–18643
54. Kamber DN, Nazarova LA, Liang Y, Lopez SA, Patterson DM, Shih H-W, Houk KN, Prescher JA (2013) J Am Chem Soc 135:13680–13683
55. Rieder U, Luedtke NW (2014) Angew Chem Int Ed 53:9168–9172
56. Niederwieser A, Späte A-K, Nguyen LD, Jüngst C, Reutter W, Wittmann V (2013) Angew Chem Int Ed 52:4265–4268
57. Engelsma SB, Willems LI, van Paaschen CE, van Kasteren SI, van der Marel GA, Overkleeft HS, Filippov DV (2014) Org Lett 16:2744–2747
58. Alley SC, Okeley NM, Senter PD (2010) Curr Opin Chem Biol 14:529–537
59. Li J, Jia S, Chen PR (2014) Nat Chem Biol 10:1003–1005
60. Michaelis J, Roloff A, Seitz O (2014) Org Biomol Chem 12:2821–2833
61. Versteegen RM, Rossin R, ten Hoeve W, Janssen HM, Robillard MS (2013) Angew Chem Int Ed 52:14112–14116
62. Li Q, Dong T, Liu X, Lei X (2013) J Am Chem Soc 135:4996–4999
63. Zhang X, Dong T, Li Q, Liu X, Li L, Chen S, Lei X (2015) ACS Chem Biol 10:1676–1683
64. Kamber DN, Liang Y, Blizzard RJ, Liu F, Mehl RA, Houk KN, Prescher JA (2015) J Am Chem Soc 137:8388–8391
65. Keliher EJ, Reiner T, Turetsky A, Hilderbrand SA, Weissleder R (2011) ChemMedChem 6:424–427
66. Devaraj NK, Thurber GM, Keliher EJ, Marinelli B, Weissleder R (2012) Proc Natl Acad Sci USA 109:4762–4767
67. Uttamapinant C, Howe JD, Lang K, Beránek V, Davis L, Mahesh M, Barry NP, Chin JW (2015) J Am Chem Soc 137:4602–4605
68. Li Z, Wang D, Li L, Pan S, Na Z, Tan CYJ, Yao SQ (2014) J Am Chem Soc 136:9990–9998
69. Schoch J, Wiessler M, Jäschke A (2010) J Am Chem Soc 132:8846–8847
70. Schoch J, Staudt M, Samanta A, Wiessler M, Jäschke A (2012) Bioconj Chem 23:1382–1386
71. Schoch J, Ameta S, Jaschke A (2011) Chem Commun 47:12536–12537
72. Asare-Okai PN, Agustin E, Fabris D, Royzen M (2014) Chem Commun 50:7844–7847
73. Pyka AM, Domnick C, Braun F, Kath-Schorr S (2014) Bioconj Chem 25:1438–1443
74. Kumar R, El-Sagheer A, Tumpane J, Lincoln P, Wilhelmsson LM, Brown T (2007) J Am Chem Soc 129:6859–6864
75. Chen XH, Roloff A, Seitz O (2012) Angew Chem 51:4479–4483
76. Sando S, Kool ET (2002) J Am Chem Soc 124:9686–9687
77. Franzini RM, Kool ET (2009) J Am Chem Soc 131:16021–16023
78. Seckute J, Yang J, Devaraj NK (2013) Nucleic Acids Res 41:e148
79. Cai J, Li X, Yue X, Taylor JS (2004) J Am Chem Soc 126:16324
80. Gorska K, Keklikoglou I, Tschulena U, Winssinger N (2011) Chem Sci 2:1969
81. Gorska K, Winssinger N (2013) Angew Chem 52:6820–6843
82. Furukawa K, Abe H, Tamura Y, Yoshimoto R, Yoshida M, Tsuneda S, Ito Y (2011) Angew Chem 50:12020–12023
83. Shibata A, Uzawa T, Nakashima Y, Ito M, Nakano Y, Shuto S, Ito Y, Abe H (2013) J Am Chem Soc 135:14172–14178
84. Lang K, Chin JW (2014) Chem Rev 114:4764–4806
85. Chalfie M, Tu Y, Euskirchen G, Ward WW, Prasher DC (1994) Science 263:802–805
86. Heim R, Prasher DC, Tsien RY (1994) Proc Natl Acad Sci 91:12501–12504
87. Shaner NC, Steinbach PA, Tsien RY (2005) Nat Methods 2:905–909
88. Keppler A, Gendreizig S, Gronemeyer T, Pick H, Vogel H, Johnsson K (2003) Nat Biotechnol 21:86–89
89. Fernandez-Suarez M, Baruah H, Martinez-Hernandez L, Xie KT, Baskin JM, Bertozzi CR, Ting AY (2007) Nat Biotechnol 25:1483–1487
90. George N, Pick H, Vogel H, Johnsson N, Johnsson K (2004) J Am Chem Soc 126:8896–8897
91. Griffin BA, Adams SR, Tsien RY (1998) Science 281:269–272
92. Halo TL, Appelbaum J, Hobert EM, Balkin DM, Schepartz A (2009) J Am Chem Soc 131:438–439
93. Noren CJ, Anthony-Cahill SJ, Griffith MC, Schultz PG (1989) Science 244:182–188
94. Neumann H, Wang K, Davis L, Garcia-Alai M, Chin JW (2010) Nature 464:441–444
95. Greiss S, Chin JW (2011) J Am Chem Soc 133:14196–14199

96. Seitchik JL, Peeler JC, Taylor MT, Blackman ML, Rhoads TW, Cooley RB, Refakis C, Fox JM, Mehl RA (2012) J Am Chem Soc 134:2898–2901
97. Lang K, Davis L, Torres-Kolbus J, Chou C, Deiters A, Chin JW (2012) Nat Chem 4:298–304
98. Lang K, Davis L, Wallace S, Mahesh M, Cox DJ, Blackman ML, Fox JM, Chin JW (2012) J Am Chem Soc 134:10317–10320
99. Plass T, Milles S, Koehler C, Szymański J, Mueller R, Wießler M, Schultz C, Lemke EA (2012) Angew Chem Int Ed 51:4166–4170
100. Kurra Y, Odoi KA, Lee Y-J, Yang Y, Lu T, Wheeler SE, Torres-Kolbus J, Deiters A, Liu WR (2014) Bioconj Chem 25:1730–1738
101. Sachdeva A, Wang K, Elliott T, Chin JW (2014) J Am Chem Soc 136:7785–7788
102. Wang K, Sachdeva A, Cox DJ, Wilf NW, Lang K, Wallace S, Mehl RA, Chin JW (2014) Nat Chem 6:393–403
103. Lukinavičius G, Umezawa K, Olivier N, Honigmann A, Yang G, Plass T, Mueller V, Reymond L, Corrêa IR Jr, Luo Z-G, Schultz C, Lemke EA, Heppenstall P, Eggeling C, Manley S, Johnsson K (2013) Nat Chem 5:132–139
104. Liu DS, Tangpeerachaikul A, Selvaraj R, Taylor MT, Fox JM, Ting AY (2012) J Am Chem Soc 134:792–795
105. Brown SP, Smith AB (2015) J Am Chem Soc 137:4034–4037
106. Devaraj NK, Upadhyay R, Haun JB, Hilderbrand SA, Weissleder R (2009) Angew Chem Int Ed 48:7013–7016
107. Karver MR, Weissleder R, Hilderbrand SA (2012) Angew Chem Int Ed 51:920–922
108. Devaraj NK, Hilderbrand S, Upadhyay R, Mazitschek R, Weissleder R (2010) Angew Chem Int Ed Engl 49:2869–2872
109. Han H-S, Devaraj NK, Lee J, Hilderbrand SA, Weissleder R, Bawendi MG (2010) J Am Chem Soc 132:7838–7839
110. Haun JB, Devaraj NK, Hilderbrand SA, Lee H, Weissleder R (2010) Nat Nano 5:660–665
111. Haun JB, Devaraj NK, Marinelli BS, Lee H, Weissleder R (2011) ACS Nano 5:3204–3213
112. Haun JB, Castro CM, Wang R, Peterson VM, Marinelli BS, Lee H, Weissleder R (2011) Scince Transl Med 3:71ra16
113. Han H-S, Niemeyer E, Huang Y, Kamoun WS, Martin JD, Bhaumik J, Chen Y, Roberge S, Cui J, Martin MR, Fukumura D, Jain RK, Bawendi MG, Duda DG (2015) Proc Natl Acad Sci 112:1350–1355
114. Zlitni A, Janzen N, Foster FS, Valliant JF (2014) Angew Chem Int Ed 53:6459–6463
115. Li Z, Cai H, Hassink M, Blackman ML, Brown RCD, Conti PS, Fox JM (2010) Chem Commun 46:8043–8045
116. Herth MM, Andersen VL, Lehel S, Madsen J, Knudsen GM, Kristensen JL (2013) Chem Commun 49:3805–3807
117. Knight JC, Richter S, Wuest M, Way JD, Wuest F (2013) Org Biomol Chem 11:3817–3825
118. Rossin R, Renart Verkerk P, van den Bosch SM, Vulders RCM, Verel I, Lub J, Robillard MS (2010) Angew Chem Int Ed 49:3375–3378
119. Evans HL, Nguyen Q-D, Carroll LS, Kaliszczak M, Twyman FJ, Spivey AC, Aboagye EO (2014) Chem Commun 50:9557–9560
120. Nichols B, Qin Z, Yang J, Vera DR, Devaraj NK (2014) Chem Commun 50:5215–5217
121. Zeglis BM, Sevak KK, Reiner T, Mohindra P, Carlin SD, Zanzonico P, Weissleder R, Lewis JS (2013) J Nucl Med 54:1389–1396
122. Hang HC, Linder ME (2011) Chem Rev 111:6341–6358
123. Erdmann RS, Takakura H, Thompson AD, Rivera-Molina F, Allgeyer ES, Bewersdorf J, Toomre D, Schepartz A (2014) Angew Chem Int Ed 53:10242–10246
124. Simon JP, Ivanov IE, Adesnik M, Sabatini DD (1996) J Cell Biol 135:355–370
125. Dube DH, Bertozzi CR (2005) Nat Rev Drug Discov 4:477–488
126. Prescher JA, Bertozzi CR (2006) Cell 126:851–854
127. Chang PV, Prescher JA, Hangauer MJ, Bertozzi CR (2007) J Am Chem Soc 129:8400–8401
128. Cole CM, Yang J, Seckute J, Devaraj NK (2013) ChemBioChem 14:205–208
129. Budin G, Yang KS, Reiner T, Weissleder R (2011) Angew Chem Int Ed 50:9378–9381
130. Yang KS, Budin G, Reiner T, Vinegoni C, Weissleder R (2012) Angew Chem Int Ed Engl 51:6598–6603
131. Su Y, Pan S, Li Z, Li L, Wu X, Hao P, Sze SK, Yao SQ (2015) Sci Rep 5:7724

Top Curr Chem (Z) (2016) 374:4
DOI 10.1007/s41061-015-0004-0

REVIEW

Cycloadditions for Studying Nucleic Acids

Stephanie Kath-Schorr[1]

Received: 28 October 2015 / Accepted: 30 November 2015 / Published online: 23 December 2015

Abstract Cycloaddition reactions for site-specific or global modification of nucleic acids have enabled the preparation of a plethora of previously inaccessible DNA and RNA constructs for structural and functional studies on naturally occurring nucleic acids, the assembly of nucleic acid nanostructures, therapeutic applications, and recently, the development of novel aptamers. In this chapter, recent progress in nucleic acid functionalization via a range of different cycloaddition (click) chemistries is presented. At first, cycloaddition/click chemistries already used for modifying nucleic acids are summarized, ranging from the well-established copper(I)-catalyzed alkyne–azide cycloaddition reaction to copper free methods, such as the strain-promoted azide–alkyne cycloaddition, tetrazole-based photoclick chemistry and the inverse electron demand Diels–Alder cycloaddition reaction between strained alkenes and tetrazine derivatives. The subsequent sections contain selected applications of nucleic acid functionalization via click chemistry; in particular, site-specific enzymatic labeling in vitro, either via DNA and RNA recognizing enzymes or by introducing unnatural base pairs modified for click reactions. Further sections report recent progress in metabolic labeling and fluorescent detection of DNA and RNA synthesis in vivo, click nucleic acid ligation, click chemistry in nanostructure assembly and click-SELEX as a novel method for the selection of aptamers.

Keywords Nucleic acid · Cycloaddition · Click chemistry · CuAAC · iEDDA · Metabolic labeling

✉ Stephanie Kath-Schorr
skath@uni-bonn.de

[1] LIMES Institute, Chemical Biology and Medicinal Chemistry Unit, University of Bonn, Bonn, Germany

1 Introduction

This chapter gives an overview about applications of cycloaddition reactions to study nucleic acids. These cycloaddition reactions are typically grouped under the heading "click chemistry" despite varying other chemistries, and an enormous amount of different applications in the context of DNA and RNA have been found so far. Only selected recent implementations for cycloaddition reactions on nucleic acids can be presented in the limited scope of this chapter. Today's click reactions, as first defined by Sharpless et al. [1], are associated with biocompatibility, fast reaction kinetics and bio-orthogonality.

Applications of cycloaddition reactions on nucleic acids span from global or site-specific labeling of modified oligonucleotides in vitro for structural and functional studies on nucleic acids, towards in vivo applications of click reactions such as metabolic DNA and RNA labeling, modification of oligonucleotides for antiviral and anti-tumor therapies and the regulation of gene expression. Moreover, cycloaddition click chemistry has been used to build up DNA and RNA strands with artificial backbones and for DNA nanostructure assembly. Examples for these applications will be given in the following sections.

All three main regions of nucleic acids have been targeted for chemical modifications that can undergo cycloaddition reactions: the nucleobases, the (deoxy-)ribose moiety and the phosphodiester backbone. The most widespread type of modifications are nucleobase alterations for the introduction of novel base analogues, bioconjugation or attaching fluorophores for structural and functional studies.

Chemically modified nucleosides can be introduced into nucleic acid by either synthetic methods during solid phase oligonucleotide synthesis or via enzymatic reactions. The site-specific functionalization of nucleic acids proceeds primarily via solid phase DNA/RNA synthesis. Artificial nucleosides that possess reactive groups for click chemistry (at the nucleobase or at sugar moiety) are introduced as phosphoramidites, either internally or at the 5′-end of the synthesized oligonucleotide. Moreover, the use of solid supports bearing modified nucleobases allows the introduction of 3′ modifications. These methods are widely employed for the site-specific attachment of labels such as fluorogenic tags, spin-labels or biotin to oligonucleotides, and are highly relevant for structural and functional studies of these nucleic acids [2, 3]. Requirements for solid-phase oligonucleotide synthesis using modified nucleobases are, first of all, the compatibility of the introduced functional group with the reaction conditions of solid-phase synthesis. Moreover, limitations in length do exist for solid-phase DNA and RNA synthesis (shorter than 100 nucleotides for RNA synthesis); these restrict investigation of site-specifically modified long oligonucleotides such as long non-coding RNA molecules with a length of several hundred nucleotides.

A different approach is the high-density functionalization of DNA and RNA via enzymatic methods. Here, modified nucleobases are introduced in vitro into oligonucleotides via enzymatic reactions, namely PCR and in vitro transcription. Various polymerases accept small-sized modifications on nucleobases, such as alkyne moieties, if the formation of Watson–Crick hydrogen bonds is not hampered by the modification [4–10]. Furthermore, end-labeling of enzymatically prepared RNA molecules can be achieved using the corresponding for post-synthetic click chemistry modified starter nucleotides for T7 in vitro transcription. For a recent example, see Jäschke et al. [11]. Here, RNA labeling was achieved using an alkyne-modified dinucleotide as starter nucleotide for in vitro transcription. The 5′-end of the transcript remains unaltered, allowing ligation to other RNA molecules and resulting in an internally modified RNA oligonucleotide [11].

In the first section of this chapter, the different cycloaddition/click chemistries already used for modifying nucleic acids are summarized. The subsequent sections contain selected applications and themes where cycloaddition reactions are a central element of studies on nucleic acids.

2 Cycloaddition Reactions on Nucleic Acids

The following graphic (Fig. 1) gives an overview about various cycloaddition reactions used for nucleic acid functionalization. The individual reactions are discussed in detail below.

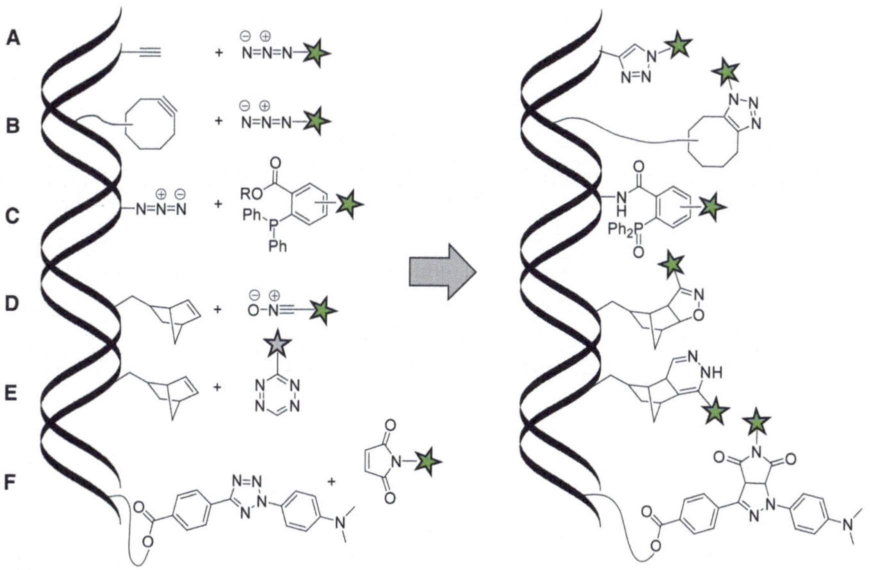

Fig. 1 Cycloaddition reactions employed in nucleic acid labeling with reporter groups (*green star*). *A* Cu^{I}-mediated azide–alkyne cycloaddition (CuAAC) of a terminal alkyne with an azide. *B* Strain-promoted azide–alkyne cycloaddition (SPAAC) of an azide with a cyclooctyne derivative. *C* Staudinger ligation of an azide with a phosphine derivative (not a cycloaddition reaction, see below). *D* Norbornene cycloaddition of a nitrile oxide as 1,3-dipole and a norbornene as dipolarophile. *E* Inverse electron-demand Diels–Alder cycloaddition reaction between a strained double bond (norbornene) and a tetrazine derivative. *F* Photo-click reaction of a push–pull-substituted diaryltetrazole with an activated double bond (maleimide)

2.1 Copper(I)-Catalyzed Alkyne–Azide Cycloaddition (CuAAC) on Nucleic Acids

The copper(I)-catalyzed alkyne–azide cycloaddition (CuAAC) reaction [1] (Fig. 1A) has been extensively employed for the site-specific labeling of oligonucleotides with various reporter groups [3, 4, 7, 9, 12–16], ligating DNA strands [17, 18], cross-linking complementary strands [19], for surface functionalization [20] and for the formation of bimetallic Ag–Au nanowires from DNA templates [21]. Selected examples are described in the following section.

Alkyne moieties can easily be incorporated into DNA and RNA strands during solid phase synthesis and the site-specific incorporation of multiple, even consecutive labels into DNA has been demonstrated [12]. Using appropriate copper(I)-complexing ligands [22] to prevent copper-catalyzed cleavage of DNA [23], azide-modified sugars and fluorophores react efficiently with alkyne-modified DNA [12]. The site-specific incorporation of different reporter groups, which is, for example, desired to study nucleic acid folding via Förster resonance energy transfer (FRET) using an appropriate FRET pair as fluorophores, can also be achieved by postsynthetic click modification. For this, alkyne functionalized nucleobases possessing orthogonal protecting groups (TMS, TIPS) are introduced as phosphoramidites during solid phase DNA synthesis [13]. Selective stepwise deprotection of alkyne moieties allows labeling of DNA oligonucleotides with up to three different reporter groups [13].

CuAAC click chemistry has further been used to cross-link complementary DNA strands to form stable duplexes employing alkyne- and azide-modified nucleobases. Increased melting temperatures up to 30 °C could be achieved [19]. PNA/DNA-peptide conjugates can easily be prepared by CuAAC click chemistry using 3′-azido-modified oligonucleotides and alkyne-modified peptides [24, 25].

Furthermore, copper catalyzed click chemistry allowed the assembly of catalytically active ribozymes with RNA strand lengths beyond the scope of solid phase RNA synthesis. In this context, the hairpin ribozyme was reconstructed from chemically synthesized RNA strands, which were covalently cross-linked by a CuAAC reaction [26]. To prevent RNA degradation, Cu^{II} sulfate and sodium ascorbate for in situ generation of the catalyst and a Cu^{I} stabilizing ligand are usually employed [22]. Azide functionalities were introduced into RNA strands via a two-step method by postsynthetically reacting amino-modified uridines with N-hydroxysuccinimide esters of an azide-modified carboxylic acid [26]. For the hammerhead ribozyme, an intrastrand click-ligation of RNA strands bearing 3′-alkyne and 5′-azide modifications gave rise to active ribozyme constructs with an artificial triazole backbone at the active site [26].

Nitroxide radical spin labels can be used to measure inter-spin distances by pulsed electron paramagnetic resonance (EPR) spectroscopy, and to allow study of structure and folding of DNA and RNA oligonucleotides [27–30]. Spin-labeling of nucleobases by CuAAC in solution has first been demonstrated using the radical azide 4-azido-2,2,6,6-tetramethylpiperidine 1-oxyl (4-azido-TEMPO) [31]. Alkyne modifications at position 7 of 7-deazapurines and C5-modified pyrimidines allowed attachment of the spin label in the major groove of duplex DNA [31]. In the same year, click chemistry on solid support to introduce spin labels into DNA was reported [32].

The incorporation of azides into DNA and RNA during solid phase synthesis has long been neglected due to incompatibility of phosphoramidite chemistry with azido groups because of Staudinger-type side-reactions [33, 34]. Instead, enzymatic methods were employed to incorporate azido-modified nucleobases as triphosphates into DNA [6, 35]. Furthermore, H-phosphonate chemistry allowed the intrastrand introduction of azide labels into short DNA oligonucleotides synthesized on solid support [36]. In 2011, Micura et al. demonstrated efficient preparation of C2′ azido-modified RNA by solid phase RNA synthesis using novel azido-modified nucleoside phosphodiester building blocks [37, 38]. All four 2′-azido–modified canonical nucleobases, 2′-azido-2′-deoxyuridine [37], 2′-azido-2′-deoxyadenosine [37], 2′-azido cytidine [38] and 2′-azido guanosine [38] were used in solid phase RNA synthesis and postsynthetic fluorescent labeling via CuAAC click chemistry using alkyne-modified fluorophores. Additionally, 2′-azido groups were well tolerated in the guide strand of siRNAs, demonstrated by siRNA induced silencing of the brain acid soluble protein 1 (BASP1) encoding gene in chicken fibroblasts [38]. In 2014, preparation of a 2′-azido–modified solid support for automated RNA synthesis compatible with phosphoramidite chemistry for 3′-terminal labeling of oligoribonucleotides was reported [39].

2.2 Copper-Free Click Chemistry for Nucleic Acid Functionalization

The presence of Cu^{1} ions leads to RNA damage and is toxic for in vivo systems. Copper-free click reactions are therefore highly desirable. Several groups established copper-free cycloaddition reactions on nucleic acids. Recent developments in the field are presented in the following sections in detail. In the same context, a modified Staudinger reaction has been employed for bioorthogonal labeling of nucleic acids; for completeness, this method is therefore also referenced briefly below.

2.2.1 Staudinger Ligation

The Staudinger ligation, adapted by Bertozzi et al. for orthogonal labeling of biomolecules [40] yields a stable amide bond by reaction of an azido-modified molecule and a triarylphosphine derivative (Fig. 1C). Incorporation of azides into nucleic acids is a prerequisite for application of this bioorthogonal labeling reaction, which is problematic during solid phase synthesis as discussed above. However, the Staudinger ligation reaction has successfully been used for DNA labeling with fluorescent reporter groups employing a 5′-azido–modified oligonucleotide [41]. Enzymatic incorporation of azido-modified 2′-deoxyuridine and 2′-deoxyadenosine triphosphates during PCR reactions allowed internal labeling of DNA strands via Staudinger ligation with biotin or fluorophore-modified phosphines [6, 42, 43]. Due to the complex synthesis of azido-modified nucleic acids, combined with the poor stability of phosphines in solution and relatively slow reaction kinetics, focus shifted more and more towards the development of other catalyst-free bioorthogonal reactions, such as the strain-promoted azide–alkyne cycloaddition discussed in the following section.

2.2.2 *Strain-Promoted Azide–Alkyne Cycloadditions*

A particular major limitation of the CuAAC reaction is metal-catalyzed strand degradation [22, 23, 44], which has been partially overcome by the development of novel Cu^{I} stabilizing ligands [45–47]. However, the cell toxicity of copper complexes remains challenging for live cell and in vivo applications. The strain-promoted azide–alkyne cycloaddition (SPAAC, Fig. 1B) between azides and strained cyclooctynes [48, 49] such as dibenzocyclooctyne allows bioorthogonal reactions on nucleic acids in the absence of copper. Solid phase RNA synthesis using a strained cyclooctyne phosphoramidite for 5′-end functionalization and reaction with azido-modified peptides and oligosaccharides [50] and 5′-end cyclooctyne modification of RNA during solid phase synthesis [51], non-nucleoside alkyne monomers for RNA labeling [52], as well as template directed DNA strand ligation [53] and internal and terminal DNA labeling with dibenzocyclooctyne [54, 55], were reported. 5′-end modified DNA with an achiral bicyclo [6.1.0] nonyne phosphoramidite during solid phase DNA synthesis can also be used for SPAAC yielding only enantiomeric products [56]. Conjugation to lipophilic polymers has been demonstrated [56].

SPAAC has further been employed for RNA labeling with 2′-azido-modified nucleotides incorporated at the 3′-end of RNA oligonucleotides by an enzymatic approach with poly(A) polymerase [7]. Ligation of these 3′-azido modified RNAs employing a splinted ligation approach allows the preparation of internally azido-modified RNAs [7]. Besides this, azido-modified capped RNA was successfully labeled via SPAAC [14].

Gold nanoparticles coated with single stranded DNA bearing a 3′-azide or 5′-alkyne modification were covalently linked together using a splint DNA strand to promote the SPAAC reaction demonstrating applications of SPAAC reactions in programmed nanoparticle organization [57].

Bibenzocyclooctyne and bicyclo [6.1.0] non-4-yne-modified 2′-deoxyuridine triphosphates can be further employed for PCR reactions, with standard DNA polymerases allowing internal labeling of DNA with multiple strained alkyne residues [10]. Fluorescent labeling was demonstrated with azide-modified fluorophores [10].

In summary, SPAAC reactions with dibenzocyclooctyne-modified nucleic acids have proven to be valuable tools for a catalyst-free bioorthogonal click reaction with biologically inert, non-natural azides. One drawback is the rather slow reaction with second order rate constants of approximately 0.05 M^{-1} s^{-1} [50, 58]. Wagenknecht et al. [59] developed a carboxymethylmonobenzocyclooctyne-modified 2′-deoxyuridine phosphoramidite which possesses increased reactivity towards azide-modified reporter groups ($k_2 = 0.8$ M^{-1} s^{-1} in aqueous solution). Incorporation into DNA during solid phase DNA synthesis allowed postsynthetic SPAAC reaction with an quinolinium-styryl-coumarin azide dye probe. Click reaction results in a strong increase in fluorescence intensity of the dye [59].

2.2.3 1,3 Dipolar Cycloadditions with Nitrile Oxides (Nitrile Oxide–Norbornene Click Chemistry)

Nitrile oxides react with strained alkenes in a Huisgen 1,3 dipolar cycloaddition reaction (Fig. 1D). This bioorthogonal reaction has been employed for labeling of DNA during automated solid phase synthesis. Nitrile oxides are strong electrophiles; however, the cross reaction with natural nucleobases is rather slow compared to the reaction with the norbornene double bond [60]. Furthermore, the reaction of styrene-modified DNA with nitrile oxides was demonstrated, yielding only one regioisomer, 3,5-disubstituted isoxazoline [5]. Styrene functionalized nucleobases were either incorporated during solid phase DNA synthesis or as triphosphates in PCR reactions of fragments up to 900 base pairs. The CuAAC click chemistry is orthogonal to the 1,3 dipolar cycloaddition, with nitrile oxides allowing high density functionalization of DNA with two different reporter groups [5].

2.2.4 Photo-Click Reaction on Nucleic Acids

Recently, tetrazole-based photoclick chemistry, which has previously been used for protein labeling in live cells [61], has been applied for postsynthetic DNA modification [62]. 2′-Deoxyuridine was modified with a substituted diaryltetrazole, bearing an electron-donating dimethylamino group at one end and an electron-withdrawing carboxyl group at the other end, and incorporated into DNA as phosphoramidite building block during solid phase DNA synthesis (Fig. 1F). Postsynthetic labeling is feasible using maleimide-modified fluorophores as dipolarophiles and irradiation at a wavelength of 365 nm [62]. The rate constant of the photo-click reaction is in the range of other fast copper-free bioorthogonal reactions such as Diels–Alder cycloaddition reactions. An advantage of this photo-click reaction is the fact that the reaction only proceeds upon irradiation. This allows an exact timing of the reaction, which might be of interest for spatiotemporal control in live cell applications [62].

2.2.5 Inverse Electron Demand Diels–Alder Cycloaddition (iEDDA) Reaction

The inverse electron demand Diels–Alder cycloaddition (iEDDA) reaction between strained alkenes and tetrazine derivatives has gained more and more attention for orthogonal labeling of biomolecules in the last years (Fig. 1E) [63]. This catalyst-free reaction can be extremely rapid in the case of *trans*-cyclooctenes as reactants with second order rate constants of up to 380,000 M^{-1} s^{-1} in aqueous solutions [64], and reactions on genetically encoded alkenes in proteins and on alkene functionalized small molecules can be performed in living cells [65–70]. The reaction is highly specific and cell toxicity of the tetrazine derivatives is negligible [66, 71, 72]. Moreover, tetrazine–fluorophore conjugates possess strong turn-on fluorescence after reaction with dienophiles that strongly depends on the fluorophore employed (fluorophore emission between 510 and 570 nm) [73–75]. This is an extremely important feature for labeling procedures in living cells and in vivo, where removal of unreacted probes is virtually impossible.

To date, a few examples of iEDDA reactions on nucleic acids in vitro and in cells were reported. In 2010, Jäschke et al. [76] reported the first example of DNA modification by the inverse-electron-demand Diels–Alder reaction between norbornene dienophiles and tetrazine-derivatives. 3′- and 5′-terminal labeling, as well as internal modification of DNA with norbornenes and subsequent iEDDA reaction, was demonstrated [76]. Yields of 96 % are observed using a 1:1 stoichiometry of DNA and tetrazine derivative [76]. Lower reactant concentrations and considerably lower excess of labeling reagent (usually 1:3 stoichiometry) compared to CuAAC, as well as the absence of any toxic catalysts, shows the potential of this cycloaddition reaction for in cell and in vivo applications.

CuAAC and iEDDA reactions are fully orthogonal and can be employed for dual labeling strategies of DNA in a one-pot reaction [64]. Facile preparation of dual labeled FRET probes for biophysical studies on nucleic acids is possible by this approach [64].

The iEDDA reaction was further employed for template-directed ligation of DNA using fluorescent DNA probes consisting of quenched tetrazine–fluorophore conjugates and methylcyclopropene groups as dienophiles [77]. This method could in principle allow fluorescent detection of specific DNA and RNA sequences in living cells.

Besides stable norbornene and slightly more reactive methylcyclopropene dienophiles, *trans*-cyclooctenes have also been incorporated into DNA during PCR as *trans*-cyclooctene–modified thymidine triphosphate and reacted with tetrazine derivatives in an iEDDA reaction [78].

The iEDDA reaction is also compatible with RNA. 5′-End labeling of an RNA modified with a 5′-norbornene phosphoramidite during solid phase RNA synthesis was performed [79]. In 2014, we demonstrated intra-strand labeling of norbornene-modified RNA oligonucleotides by iEDDA in vitro and in a cellular context [80]. For this, a norbornene-modified uridine phosphoramidite was introduced into RNA during solid phase RNA synthesis. Transfection of norbornene containing siRNA duplexes into mammalian cells and subsequent reaction with tetrazine–fluorophore derivatives in the cytoplasm of cells proved that the iEDDA reaction can in principle be used to label nucleic acids in living cells [80].

RNA-peptide conjugates have also been prepared by the inverse electron demand Diels–Alder reaction using tetrazine-modified peptides and enzymatically 5′-dienophile–modified RNA [81]. Moreover, enzymatic RNA labeling using *trans*-cyclooctene–modified cytidine triphosphate during in vitro transcription allows the addition of reporter groups via iEDDA at internal sites of the target RNA [82].

3 Click Chemistry for Detection of Natural Oligonucleotides

The detection of specific, endogenous DNA and RNA sequences, especially in living cells, is challenging and can provide valuable information about their localization, expression levels and their lifetime. Being able to image specific DNA and RNA molecules in vivo in real-time could offer plenty of information on oligonucleotide synthesis, processing, transport and RNA degradation, with new insights for drug discovery and medical diagnostics. Various approaches aim to label endogenous DNA and RNA with reporter groups such as fluorophores for detection. In this section, methods for labeling specific, unmodified DNA and RNA sequences in vitro (with potential for future in cell labeling) using click chemistry are briefly presented.

3.1 Site-Specific Labeling via DNA and RNA Recognizing Enzymes

DNA-specific enzymes, DNA methyltransferases (Mtases) can be employed to transfer activated groups sequence specifically to DNA from a cofactor, usually from modified *S*-adenosylmethionine (AdoMet) analogues. This approach has been used as a two-step labeling approach by first alkylating plasmid DNA via an alkyne derivatized AdoMet at the N^6-amino position of adenine and subsequent CuAAC reaction with an azide-modified fluorophore (Fig. 2) [83]. Three DNA methyltransferases belonging to the class of adenine-N^6 methyltransferases were identified from a screening of wild-type methyltransferases enzymes to successfully catalyze the alkyltransfer reaction: M.TaqI (target sequence 5′-TCGA-3′), M.XbaI (5′-TCTAGA-3′) and M.FokI (5′-GGATG-3′ and 5′-CATCC-3′) [83]. A major drawback of this approach is the fragmentation of the plasmid DNA during the CuAAC reaction, despite the presence of Cu^I stabilizing ligands [83]. Copper-free click reactions such as SPAAC and iEDDA will certainly solve this issue.

A similar chemo-enzymatic approach has been used for fluorescent labeling of tRNA in vitro [84]. The tRNA-specific methyltransferase Trm1 was employed to transfer a pent-2-en-4-ynyl group from a modified AdoMet analogue to the N^2 position of guanosine 26 in $tRNA^{Phe}$ for subsequent CuAAC labeling [84].

Using the sequence-guided 2′-*O*-methyltransferase in box C/D ribonucleoprotein particles, sequence-specific RNA labeling via a two-step approach involving CuAAC click chemistry could also be achieved in vitro [85]. Notable for this approach is that the sequence-specificity is generated by the use of "guide RNAs," which allows labeling of various different RNA sequences by altering the guide sequence [85]. Despite the need for a complementary guide sequence to define the target RNA sequence, this method might in future show great potential for in cell labeling of RNA.

Rentmeister et al. developed a two-step approach to site-specifically label the 5′-cap of eukaryotic mRNAs via CuAAC [15] and SPAAC [14] click chemistry. For this, a variant of trimethylguanosine synthase 2 from *Giardia lamblia* is employed,

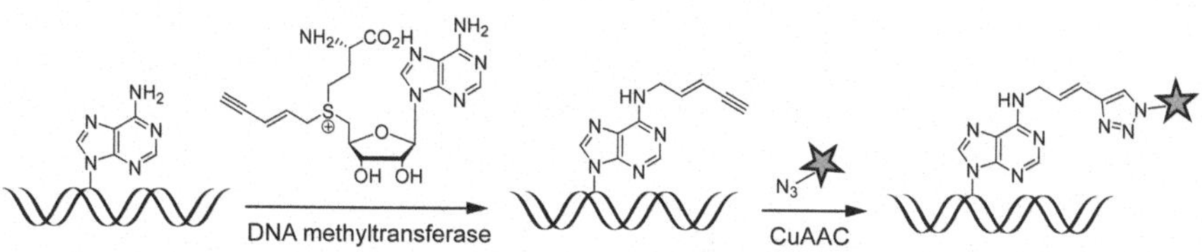

Fig. 2 Methyltransferase-directed N^6-adenine alkylation of DNA by 5′-[(*S*)-[(3*S*)-3-amino-3-carboxylpropyl](*E*)-pent-2-en-4-ynylsulfonio]-5′-deoxy-adenosine and subsequent CuAAC reaction with an azido-modified reporter group [83]

which recognizes the m^7G-cap structure of mRNA and transfers an alkyne group from a modified *S*-adenosylmethionine (AdoMet) analogue [15]. Further engineering of the enzyme allowed terminal azido-labeling of the 5′-cap for strain-promoted azide–alkyne cycloaddition (SPAAC) [14]. Recently, this method was extended to the transfer of a 4-vinylbenzyl group for inverse electron-demand Diels–Alder and photo-click reactions using a novel AdoMet-analogue [86]. The two-step labeling approach was demonstrated in eukaryotic cell lysate and in the future might be applicable for mRNA labeling in living cells, once complex issues, such as the synthesis of AdoMet analogues by cells and specificity of engineered enzymes for these analogues, have been solved.

3.2 Site-Specific RNA Labeling Using DNAzymes

Enzymatically catalyzed transfer of bioorthogonal groups to DNA and RNA is not solely limited to protein enzymes. Höbartner et al. [87] developed an in vitro approach for site-specific post-transcriptional RNA labeling using a deoxyribozyme (DNAzyme). Guanosine derivatives bearing bioorthogonal groups such as azide moieties were site-specifically transferred to the 2′-OH group of adenines in RNA oligonucleotides [88]. The DNAzyme recognizes the target nucleoside in the RNA substrate by hybridization. In the presence of Tb^{3+} as cofactor, a 2′,5′-phosphodiester bond is formed between the 2′-hydroxyl group of the adenosine in the target RNA and the α-phosphate of a 2′-azido–modified guanosine triphosphate derivative. Subsequent CuAAC click reaction allows the attachment of reporter groups such as fluorophores [88]. Long RNA oligonucleotides of up to 160 nucleotides are selectively labeled by this method [88]. This novel approach will certainly prove very useful for the selective labeling of long in vitro transcribed RNAs for biophysical studies.

4 Backbone Modifications and Click Nucleic Acid Ligation

Click chemistry has also been used to synthesize artificial DNA and RNA backbones via triazole linkages [89]. The first examples of an unnatural, biocompatible DNA backbone assembled by CuAAC click chemistry were demonstrated by Nakamura et al. [90] and Brown et al. [91]; the latter named this method “click nucleic acid ligation” [53, 89, 92]. PCR amplification of a DNA oligonucleotide linked via 3′-azido-dT and 5′-propargylamido-dT by CuAAC was demonstrated by Brown’s group in 2009; however, only the incorporation of a single thymidine residue was revealed by sequencing due to misreading of the triazole backbone by the DNA polymerase [91]. A more flexible second generation triazole linkage retaining the 3′-oxygen and more accurately mimicking the phosphate backbone could correctly be read by DNA polymerases. Both amplification in vitro by PCR as well as replication of a plasmid containing the triazole linkage in *E. coli* were demonstrated [93]. This approach was further extended to assemble long DNA oligonucleotides by multiple sequential ligation employing a masked azide approach and CuAAC and SPAAC reactions on solid phase [92]. The biocompatibility of the triazole linkage was further investigated in mammalian cells using a click-linked gene encoding the fluorescent protein mCherry [94]. Error-free transcription of the click-linked gene was demonstrated [94].

For RNA, artificial self-cleaving hammerhead ribozymes with a triazole linkage at the active site are reported [26]. Interestingly, triazole backbone modifications in RNA by CuAAC and SPAAC chemical ligation of 3′-azide and 5′-cyclooctyne oligonucleotides can further be reverse transcribed into DNA while one nucleotide is omitted at the linkage [95].

In summary, click ligation of nucleic acids allows the assembly of long DNA or RNA strand with various reporter groups, which are, for example, incompatible with enzymatic ligation conditions or if an efficient enzymatic ligation is prevented by extensive secondary structures of the nucleic acid.

5 Unnatural Base Pairs for Enzymatic Site-Specific Labeling of Oligonucleotides Using Click Chemistry

Incorporation of site-specific anchor groups for the attachment of labels such as fluorogenic tags, spin labels or biotin is highly desirable for structural and functional studies on oligonucleotides, such as large, naturally occurring RNA molecules with usually complex foldings. These anchor groups need to be positioned at specific sites within the sequence, where they do not impede folding and function of the RNA molecule. However, this poses a problem for RNAs with sequences longer than 100 nucleotides, due to the limitations in solid phase RNA synthesis [3]. Complex ligation strategies are required to assembly RNA molecules with a length of 200–300 bases [96], a common size for various naturally occurring non-coding RNAs [97, 98], such as ribozymes or riboswitches. Ligation of highly structured RNAs often fails or results in low yields.

An enzymatic approach for site-specific incorporation of modified triphosphates overcomes this obstacle: Unnatural base pairs can be employed for the site-specific introduction of various functional groups into DNA and RNA via standard enzymatic reactions such as PCR and in vitro transcription. Several research groups have been concentrating on the development of a third, unnatural base pair over the last decade which can be PCR amplified and transcribed into RNA in vitro by standard polymerases with almost similar efficiency and fidelity compared to the natural base pairs [99–112].

A particular focus of interest lies on so-called hydrophobic base pairing systems (UBPs), which rely on shape complementary rather than hydrogen bonding interactions of the pairing bases, as first proposed by Kool et al. [113–115]. To date, different optimized hydrophobic base pairing systems exist, developed by the groups of Romesberg [107, 111, 112, 116–123] and Hirao [105, 106, 109, 110, 124–133]. Efficient PCR amplification of these base pairs was demonstrated; furthermore, in vitro transcription into RNA using the corresponding ribose building blocks has been reported [125, 127, 133–144]. Direct functionalization of these unnatural bases with biotin or fluorophores allows site-specific modification of PCR amplified DNA [143] and transcribed RNA oligonucleotides [136, 140, 145]. However, direct introduction of functional groups is limited by their size and shape and unnatural bases, bearing, for example, large fluorgenic tags are poorly incorporated by polymerases [143, 146].

Hirao et al. [136] demonstrated in 2005 that post-transcriptional labeling of a short 17mer RNA fragment containing an amino-modified unnatural 2-oxo-(1H)pyridine base is feasible. The unnatural base was site-specifically introduced via enzymatic in vitro transcription and subsequently reacted with the N-hydroxysuccinimidylester of 5-carboxyfluorescein or 5-carboxytetramethylrhodamine.

Analogously, the d5SICS–dNaM unnatural base pair developed by Romesberg et al. [107, 142] tolerates amino-modifications at both components of this unnatural base pair with sufficient efficiency in in vitro transcription reactions to allow post-synthetic RNA labeling using NHS-ester chemistry [111, 143]. As a first example of post-transcriptional functionalization of a long RNA, a site-specifically labeled 77 nucleotide amber suppressor tyrosyl tRNA from *Methanocaldococcus jannaschii* was synthesized via in vitro transcription and post-transcriptionally reacted with NHS-biotin [143].

5.1 Site-Specific DNA Functionalization with Alkyne Moieties

Functionalization of these unnatural nucleobases by 'clickable' moieties that are relatively small in size allows efficient incorporation during PCR and transcription, and subsequent attachment of various reporter groups. This versatile approach is particularly attractive for enzymatic synthesis of long DNA and RNA strands with site-specific modifications. Post-amplificational and transcriptional labeling via click reactions is more efficient, robust and particularly less pH-dependent compared to NHS-ester chemistry. Click functionalization of DNA strands containing unnatural base pairs was demonstrated by Romesberg et al. [120]. Amongst others, the d5SICS^{CO}–dNaM unnatural base pair (Fig. 3a) was employed to introduce an alkene moiety into PCR amplified DNA oligonucleotides at one and two specific positions. An azide-modified biotin tag as well as two molecules of the N-terminal Src homology 3 domain from the human CrkII adaptor protein (nSH3) were attached via Cu^{I}-mediated azide–alkyne click reaction after PCR amplification of the 133-nucleotide-long DNA fragment.

5.2 Site-Specific RNA Functionalization with Alkynes, Azides and Strained Alkenes

Site-specific RNA labeling using an alkyne-modified hydrophobic UBP termed Ds–Pa pair [125, 138, 140, 141] was first demonstrated by Hirao et al. [135]. The alkyne linker was attached to the Pa nucleobase for postsynthetic RNA labeling using copper-catalyzed azide–alkyne click chemistry (Fig. 3b, 1) [135], resulting in a versatile labeling method for large RNA molecules. Post-transcriptional attachment of large reporter groups such as fluorophores avoids reduced transcription efficiency and selectivity caused by these bulky functional groups [146]. By this method, a 75-nucleotide tRNA molecule was site-specifically alkyne-functionalized at position 33 and subsequently reacted with fluorescent azides or azide-modified biotin [135].

Introduction of strained unsaturated ring systems such as norbornene to an UBP allows bio-orthogonal inverse electron-demand Diels–Alder cycloadditions (iEDDA) [63, 66–69, 71–73, 75, 78, 80, 86, 147–150] on RNA with tetrazine

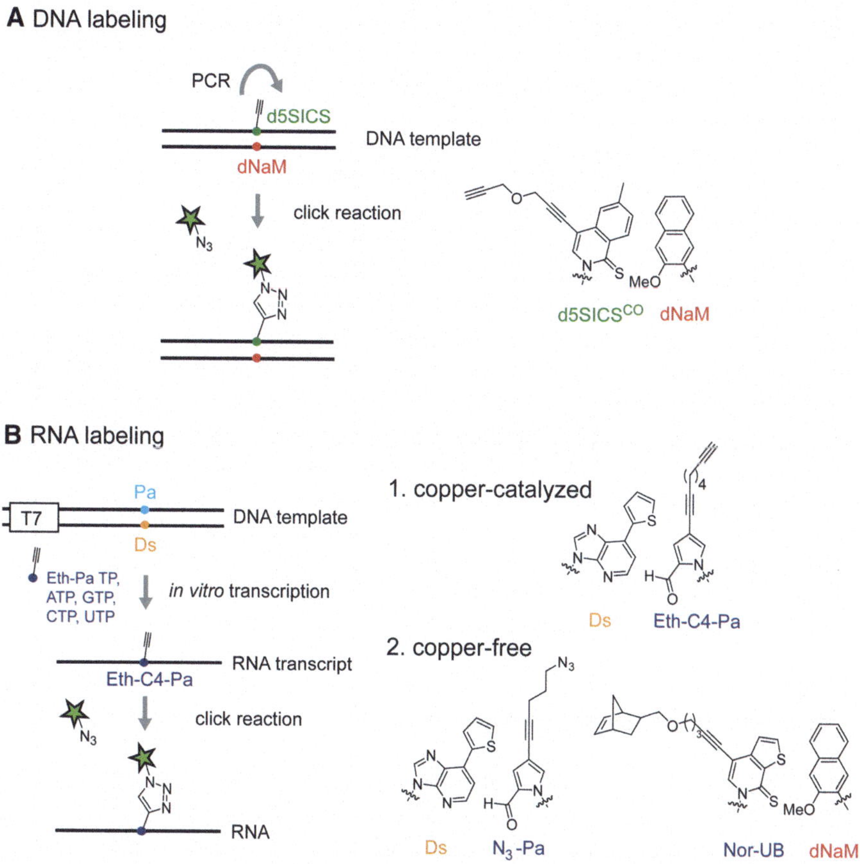

Fig. 3 Unnatural base pairs for enzymatic site-specific labeling of oligonucleotides using click chemistry. **a** DNA labeling using the unnatural d5SICS–dNaM base pair [120]. PCR amplification was carried out with an alkyne-modified d5SICS nucleoside that can be reacted with azide containing compounds. **b** RNA labeling via T7 in vitro transcription. The functionalized unnatural triphosphate is site-specifically incorporated into an RNA transcript in an in vitro transcription reaction and post-transcriptionally labeled via click chemistry. Triphosphates of the depicted unnatural bases for copper-catalyzed (Eth–C_4–Pa [135]) and copper-free click reactions (N_3–Pa [152] and Nor-UB [144]) have been developed

derivatives using non-denaturing conditions during the postsynthetic labeling step while maintaining fast reaction kinetics [66, 67, 80, 144, 150]. We modified one component of an unnatural base pair system developed by Romesberg et al. to introduce one or two norbornene moieties at predefined positions into RNA oligonucleotides in an in vitro transcription reaction using the RNA triphosphate Nor-UB and the unnatural base dNaM as template (Fig. 3b, 2). This allows functionalization of these RNA molecules directly after transcription with tetrazine derivatives containing for instance fluorophores or biotin. The native structure of the RNA will not be affected by the mild click reaction conditions, which can even be realized in a cellular context [66, 67, 80, 150]. This is of particular interest for labeling RNAs with complex folding pathways, such as riboswitches, ribozymes or aptamers, which cannot be correctly refolded in vitro after a denaturing purification step [151].

Likewise, Hirao et al. applied copper free click chemistry for RNA functionalization via the 7-(2-thienyl)imidazo[4,5-b]pyridine (Ds) and pyrrole-2-carbaldehyde (Pa) unnatural base pair [152]. For this, an azide-modified Pa nucleobase (N_3–Pa) was incorporated as triphosphate in a T7 in vitro transcription reaction using Ds as corresponding nucleobase in the DNA template (Fig. 3b, 2). Post-transcriptional modification was achieved using fluorescent dibenzocyclooctyne (DIBO) derivatives in a strain-promoted azide–alkyne cycloaddition reaction. Transcription and efficient site-specific labeling of a 260mer RNA was demonstrated.

In summary, copper-catalyzed and copper-free click chemistry approaches combined with enzymatic site-specific introduction of reactive groups provide a powerful tool for labeling of long DNA and RNA molecules that cannot be synthesized via solid-phase synthesis. The concept of unnatural base pairs is no longer limited to in vitro applications. Romesberg et al. [112] demonstrated the faithful replication of a plasmid containing the d5SICS–dNAM base pair in *E. coli*. In principle, the site-specific functionalization of target RNA molecules expressed in a cell could be feasible in the future by using one of the described triphosphates. This would be an innovative and extremely useful tool for in vivo imaging of RNA, and could enable new insights into localization and transport of specific RNA molecules.

6 Metabolic Labeling of DNA and RNA

Copper-catalyzed and copper-free click chemistries have been used for metabolic labeling and fluorescent detection of DNA and RNA synthesis in vivo. Alkyne or alkene modifications on natural nucleosides are sufficiently small to allow efficient incorporation into DNA/RNA by cellular polymerases and circumvent drawbacks of immunohistochemical staining methods using 5-bromo-2′-deoxyuridine (Br-dU), such as the poor permeability of antibodies into various tissues.

A first example for metabolic labeling of DNA was the incorporation of 5-ethynyl-2′-deoxyuridine [12] (EdU, Fig. 4A) into cellular DNA and subsequent Cu^{I}-mediated azide–alkyne click reaction using fluorescent azides. [153] The labeling process of genomes of replicating cells is schematically described in Fig. 4B. Components for the EdU labeling strategy have been commercially available since 2008 to study cell proliferation and differentiation. In contrast to Br-dU labeling, no prior nucleic acid denaturation by acid, heat or DNase digestion to allow anti-Br-dU antibody binding is required for detection. The harsh conditions required for anti-Br-dU antibody detection can, for example, destroy tissue structures. The small sized fluorescent azides can, in contrast to antibodies, diffuse through tissues and react with alkyne moieties in the genome without DNA denaturation. For EdU labeling, cells are incubated with EdU followed by fixation and detergent permeabilization prior to the addition of the staining mixture containing the fluorescent azide $CuSO_4$ and sodium ascorbate [153]. The method has also been used for labeling in whole animals; for example, for studying cell proliferation in the nervous system of embryonic [154] and adult [155] mice.

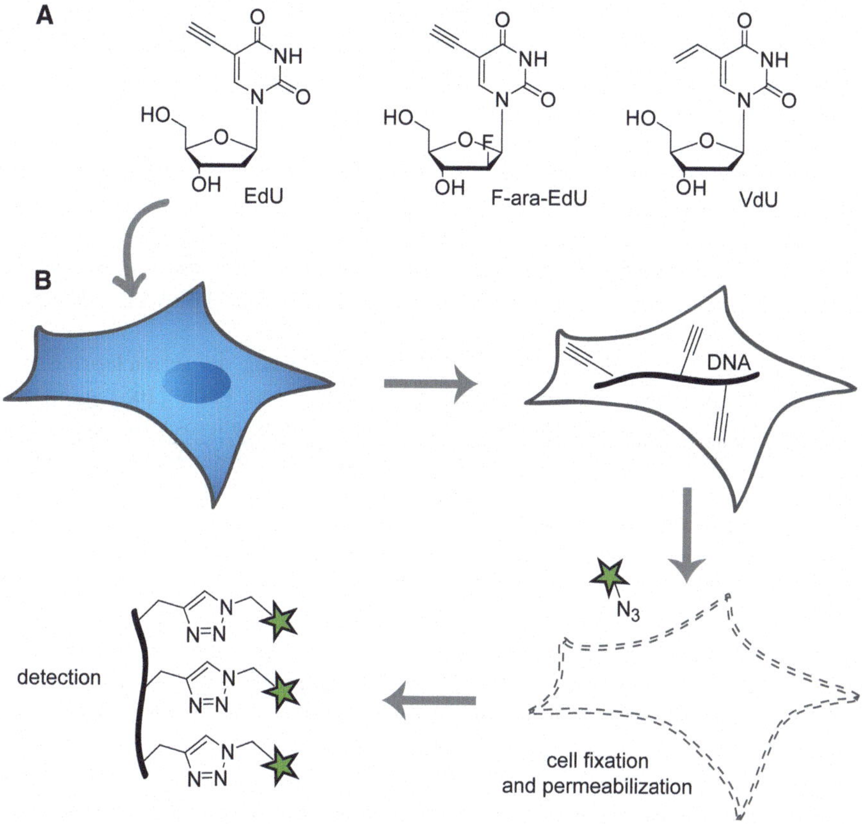

Fig. 4 Metabolic DNA labeling using alkyne- or alkene-modified nucleosides. *A* 5-Ethynyl-2′-deoxyuridine (EdU) [153], (2′*S*)-2′-deoxy-2′-fluoro-5-ethynyluridine (F-ara-EdU) [160] and 5-vinyl-2′-deoxyuridine (VdU) [167]. *B* Labeling strategy: cells are incubated with the alkyne-modified nucleoside EdU, which is incorporated into the cellular genome in proliferating cells. Subsequently, the cells are fixed, permeabilized and components for the Cu^I-mediated azide–alkyne click reaction with fluorescent azides are added. Detection can be carried out by flow cytometry or fluorescence microscopy

A similar strategy is applied for monitoring de novo RNA synthesis in cells and even in organs of whole animals using the corresponding ribose derivative, 5-ethynyl-uridine (EU) [156]. Significant labeling of cellular DNA is not observed, allowing RNA detection with high sensitivity using fluorescent azides.

A major drawback of metabolic labeling strategies using EdU is the cytotoxicity of EdU, which leads to DNA damage, cell cycle arrest and apoptosis [157–159]. Thus, the application of EdU in long-term labeling experiments is limited. Other alkyne-modified nucleosides have been developed and used in metabolic labeling experiments that exhibit lower cytotoxic effects. Interestingly, the D-arabinose configurated (2′*S*)-2′-deoxy-2′-fluoro-5-ethynyluridine (F-ara-EdU) (Fig. 4A) could be incoporated into cellular DNA with high efficiency, and possesses minimal impact on cellular proliferation [160]. Zebrafish embryogenesis was investigated over multiple days, demonstrating that successful long-term labeling is possible due to the lowered cytotoxicity of F-ara-EdU [160]. Alkyne-modified cytidine (EdC [161, 162]), adenine (EdA [163]) and guanine (EdG [163]) derivatives with slightly reduced cytotoxicity compared to EdU have also been employed for metabolic DNA labeling in cells and zebrafish embryos and could be useful tools to study pyrimidine and purine nucleoside metabolism.

The ribose equivalent 5-ethynylcytidine (EC) proved to be a valuable tool for monitoring RNA synthesis in vitro and in vivo (mice), exhibiting similar sensitivity to EU and was faster metabolized than EU [8]. Interestingly, RNA labeling with the purine nucleoside 5-ethynyl-adenosine (EA) in contrast did result in unspecific labeling of both cytoplasm and nuclei, and is not suitable for the specific detection of RNA synthesis in cells [8].

EdC, F-ara-EdU and EdA, as well as the corresponding ribose nucleoside EU, have also been used for tracking viral DNA [164] and RNA [165] genomes, respectively, in host cells. The use of super-resolution microscopy techniques allowed detection at single viral genome sensitivity [164]. Pathogen specific DNA labeling of cells infected with Herpes Simplex Virus-1 (HSV-1) has recently been achieved with an alkyne-modified gemcitabine metabolite 2′-deoxy-2′,2′-difluoro-5-ethynyluridine [166]. This nucleoside derivative is metabolized by a viral low fidelity thymidine kinase resulting selective labeling of HSV-1 infected cells.

A disadvantage of all metabolic labeling strategies based on azide–alkyne click reactions is the required Cu(I) for catalysis, which degrades cellular DNA and RNA. In 2014, metabolic labeling of DNA could be achieved using the catalyst-free inverse electron demand Diels–Alder cycloaddition reaction between alkenes and tetrazines [167]. 5-Vinyl-2′-deoxyuridine (VdU, Fig. 4A) was metabolically incorporated into genomic DNA of HeLa cells and subsequently visualized with a fluorescent dipyridyl tetrazine derivative. Its genotoxicity is reduced compared to EdU [167]. Orthogonal labeling using VdU in combination with EdU/F-ara-dU and Br-dU, followed by staining with a tetrazine–fluorophore conjugate, a fluorescent azide and the Br-dU antibody is possible, allowing multi-color labeling to observe spatial and temporal distribution of genomic DNA during *S*-phase [167].

An azide-modified nucleoside has also been used for copper-free DNA detection with fluorescent cyclooctynes via strain-promoted azide–alkyne cycloaddition. 5-(Azidomethyl)-2′-deoxyuridine (AmdU) is, in contrast to various other azide containing nucleosides, stable in solution and metabolically stable and could be incorporated in cellular DNA [35]. Recently, this method has been extended for the detection of cellular RNA by 5-azidopropyl-modified UTP analogs [168].

Nucleosides modified with strained alkenes such as cyclopropenes, norbornenes or cyclooctenes have not been tested for metabolic labeling so far. These alkenes exhibit faster reaction rates with tetrazines than VdU; however, incorporation of the modified nucleosides into DNA by cellular enzymes might be limited due to the sterically demanding substituents. If successful, live-cell imaging of cellular DNA and RNA may be feasible in the future, using such nucleosides modified for fast iEDDA cycloaddition reactions with tetrazine derivatives. For this, tetrazine–fluorophore conjugates that exhibit significant turn-on fluorescence upon reaction with alkenes (usually with fluorophores emitting between 510 and 570 nm) have to be employed, to avoid strong background fluorescence.

7 Click-SELEX

In vitro selection (SELEX, Systematic Evolution of Ligands by Exponential Enrichment) has been employed for over a decade to evolve aptamers and artificial ribozymes with various catalytic functions [169, 170]. However, nucleic acids are, in contrast to proteins, limited to the four nucleobases, thus occupying a more narrow chemical space, which can limit successful SELEX experiments against particular targets. Modified nucleosides in nucleic acid libraries are useful tools in aptamer SELEX to expand the chemical space of DNA [171–174].

The introduction of functional groups on canonical nucleobases can be restrained, because these modified nucleobases need to be compatible with the enzymatic steps required in the SELEX process. A novel approach termed click-SELEX uses a modular strategy based on Cu^{I}-catalyzed azide–alkyne cycloaddition to generate modified nucleic acid libraries (Fig. 5) [175].

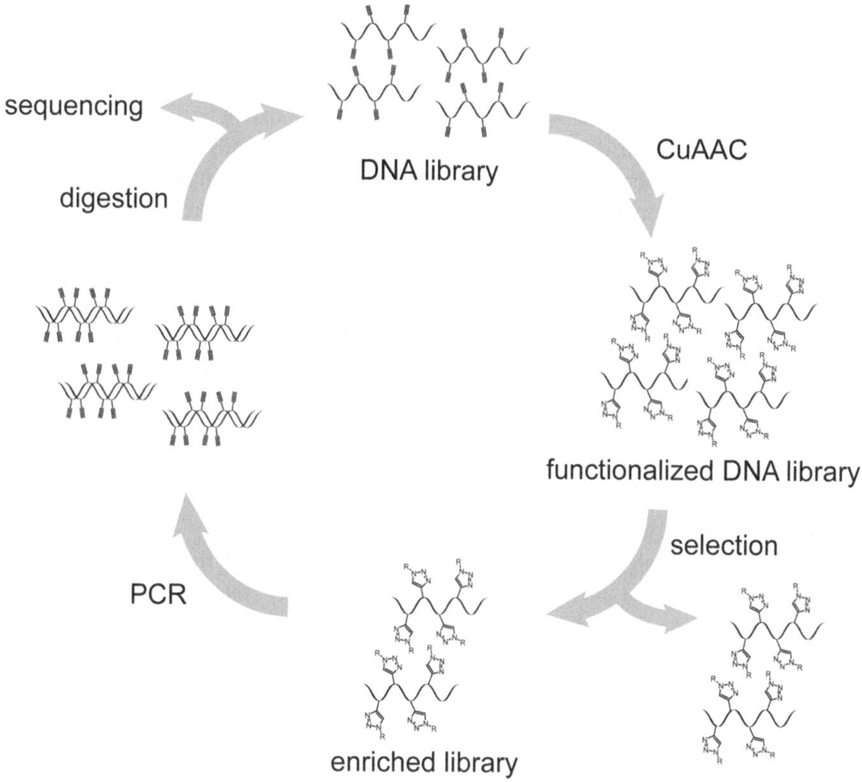

Fig. 5 Schematic representation of the click-SELEX process: An EdU containing DNA library is functionalized via CuAAC using an azide-modified molecule. The library is incubated with the target molecule, unbound sequences are removed and the enriched library is amplified by PCR using EdU triphosphate instead of the canonical thymidine to reintroduce the alkyne moiety. Single-stranded DNA is prepared by λ-exonuclease digestion. The modification can again be introduced by CuAAC for the next selection cycle. Adapted from [175]

Alkyne-modified libraries containing 5-ethynyl-2′-deoxyuridine [12] (EdU, Fig. 4A) instead of thymidine are prepared. Functionalization of the library is achieved by copper-catalyzed click reaction with azide containing compounds. As a proof of principle, 3-(2-azidoethyl)indole was used and successful selection was performed against cycle 3 GFP [175]. The amplification step after selection is carried out in the presence of EdU to restore an enriched alkyne-modified library for subsequent rounds of selection. At this point, correct amplification of the functionalized template strand by DNA polymerases is still a requirement for successful selection, and can restrict the size of the azide-modified compounds used.

In future, this might be achieved by the introduction of photocleavable linker molecules to remove large residues prior to PCR amplification [175]. Thus, various chemical moieties such as aromatic residues, amino acids or lipid modifications could be introduced in nucleic acid libraries by this method, allowing rapid access to modified libraries for aptamer selection and the selection of novel DNA catalysts.

8 Click Chemistry in Nanostructure Assembly

Click chemistry has proven to be a valuable tool to stabilize self-assembled nucleic acid nanostructures by covalent cross-links that are more stable to denaturing agents, or can, for example, be freeze-dried. Discussing various aspects of nanostructure stabilization will go beyond the scope of this chapter. Thus, only selected examples involving click chemistry for DNA nanostructure assembly and stabilization are presented.

Hexagonal DNA modules have been self-assembled and stabilized via CuAAC click chemistry by six simultaneous reactions, a method which is applicable for stabilization of various DNA nanostructures [176].

DNA nanopatterns on surfaces have been immobilized using click chemistry [177], and branched, Y-shaped DNA molecules can be prepared from tripropargylated oligonucleotides [178] by CuAAC click reactions, which are useful building blocks for higher DNA nanostructures. Moreover, SPAAC click chemistry in combination with orthogonal photochemical fixation has been used to synthesize oligomeric DNA scaffolds from cyclic DNA nanostructures which are stable towards denaturation and allow facile purification [179].

In a recent example, Cassinelli et al. [180] demonstrated that highly complex DNA-catenanes can be formed from 24 simultaneously interlocked DNA rings, resembling a chainmail architecture. Ring closure was achieved via CuAAC click chemistry [180]. Applications of this approach in nanobioelectronics, nanooptics or nanomedicine are imaginable.

9 Conclusion and Perspectives

In summary, click chemistry on nucleic acid has revolutionized the synthesis and assembly of chemically modified oligonucleotides for various biophysical and biochemical applications. In vitro preparation of labeled DNA and RNA via copper-catalyzed and copper-free click chemistry is well established to date.

Future applications might include progress in selection of functional nucleic acids (aptamers, ribozymes) with chemical entities not naturally present in DNA (or RNA), the preparation of even more complex nanomaterials built from nucleic acids, and certainly, progress in the development of novel approaches for site-specific labeling of nucleic acids in living cells or organisms to study the spatiotemporal distribution of specific DNA and RNA molecules.

References

1. Kolb HC, Finn MG, Sharpless KB (2001) Click chemistry: diverse chemical function from a few good reactions. Angew Chem 40(11):2004–2021
2. El-Sagheer AH, Brown T (2010) Click chemistry with DNA. Chem Soc Rev 39(4):1388–1405. doi:10.1039/b901971p
3. Paredes E, Das SR (2011) Click chemistry for rapid labeling and ligation of RNA. Chembiochem 12(1):125–131. doi:10.1002/cbic.201000466
4. Gramlich PM, Wirges CT, Gierlich J, Carell T (2008) Synthesis of modified DNA by PCR with alkyne-bearing purines followed by a click reaction. Org Lett 10(2):249–251. doi:10.1021/ol7026015
5. Gutsmiedl K, Fazio D, Carell T (2010) High-density DNA functionalization by a combination of Cu-catalyzed and Cu-free click chemistry. Chemistry 16(23):6877–6883. doi:10.1002/chem.201000363
6. Weisbrod SH, Marx A (2007) A nucleoside triphosphate for site-specific labelling of DNA by the Staudinger ligation. Chem Commun 18:1828–1830. doi:10.1039/b618257g
7. Winz ML, Samanta A, Benzinger D, Jäschke A (2012) Site-specific terminal and internal labeling of RNA by poly(A) polymerase tailing and copper-catalyzed or copper-free strain-promoted click chemistry. Nucleic Acids Res 40(10):e78. doi:10.1093/nar/gks062
8. Qu D, Zhou L, Wang W, Wang Z, Wang G, Chi W, Zhang B (2013) 5-Ethynylcytidine as a new agent for detecting RNA synthesis in live cells by “click” chemistry. Anal Biochem 434(1):128–135. doi:10.1016/j.ab.2012.11.023
9. Wenge U, Ehrenschwender T, Wagenknecht HA (2013) Synthesis of 2′-*O*-propargyl nucleoside triphosphates for enzymatic oligonucleotide preparation and “click” modification of DNA with Nile red as fluorescent probe. Bioconjug Chem 24(3):301–304. doi:10.1021/bc300624m
10. Ren X, Gerowska M, El-Sagheer AH, Brown T (2014) Enzymatic incorporation and fluorescent labelling of cyclooctyne-modified deoxyuridine triphosphates in DNA. Bioorg Med Chem 22(16):4384–4390. doi:10.1016/j.bmc.2014.05.050
11. Samanta A, Krause A, Jäschke A (2014) A modified dinucleotide for site-specific RNA-labelling by transcription priming and click chemistry. Chem Commun 50(11):1313–1316. doi:10.1039/c3cc46132g
12. Gierlich J, Burley GA, Gramlich PM, Hammond DM, Carell T (2006) Click chemistry as a reliable method for the high-density postsynthetic functionalization of alkyne-modified DNA. Org Lett 8(17):3639–3642. doi:10.1021/ol0610946
13. Gramlich PM, Warncke S, Gierlich J, Carell T (2008) Click–click–click: single to triple modification of DNA. Angew Chem 47(18):3442–3444. doi:10.1002/anie.200705664
14. Holstein JM, Schulz D, Rentmeister A (2014) Bioorthogonal site-specific labeling of the 5′-cap structure in eukaryotic mRNAs. Chem Commun 50(34):4478–4481. doi:10.1039/c4cc01549e
15. Schulz D, Holstein JM, Rentmeister A (2013) A chemo-enzymatic approach for site-specific modification of the RNA cap. Angew Chem 52(30):7874–7878. doi:10.1002/anie.201302874
16. Seidu-Larry S, Krieg B, Hirsch M, Helm M, Domingo O (2012) A modified guanosine phosphoramidite for click functionalization of RNA on the sugar edge. Chem Commun 48(89):11014–11016. doi:10.1039/c2cc34015a
17. Kumar R, El-Sagheer A, Tumpane J, Lincoln P, Wilhelmsson LM, Brown T (2007) Template-directed oligonucleotide strand ligation, covalent intramolecular DNA circularization and catenation using click chemistry. J Am Chem Soc 129(21):6859–6864. doi:10.1021/ja070273v

18. El-Sagheer AH, Kumar R, Findlow S, Werner JM, Lane AN, Brown T (2008) A very stable cyclic DNA miniduplex with just two base pairs. Chembiochem 9(1):50–52. doi:10.1002/cbic.200700538
19. Kocalka P, El-Sagheer AH, Brown T (2008) Rapid and efficient DNA strand cross-linking by click chemistry. Chembiochem 9(8):1280–1285. doi:10.1002/cbic.200800006
20. Rozkiewicz DI, Gierlich J, Burley GA, Gutsmiedl K, Carell T, Ravoo BJ, Reinhoudt DN (2007) Transfer printing of DNA by "click" chemistry. Chembiochem 8(16):1997–2002. doi:10.1002/cbic.200700402
21. Fischler M, Simon U, Nir H, Eichen Y, Burley GA, Gierlich J, Gramlich PM, Carell T (2007) Formation of bimetallic Ag–Au nanowires by metallization of artificial DNA duplexes. Small 3(6):1049–1055. doi:10.1002/smll.200600534
22. Chan TR, Hilgraf R, Sharpless KB, Fokin VV (2004) Polytriazoles as copper(I)-stabilizing ligands in catalysis. Org Lett 6(17):2853–2855. doi:10.1021/ol0493094
23. Thyagarajan S, Murthy NN, Narducci Sarjeant AA, Karlin KD, Rokita SE (2006) Selective DNA strand scission with binuclear copper complexes: implications for an active Cu_2–O_2 species. J Am Chem Soc 128(21):7003–7008. doi:10.1021/ja061014t
24. Gogoi K, Mane MV, Kunte SS, Kumar VA (2007) A versatile method for the preparation of conjugates of peptides with DNA/PNA/analog by employing chemo-selective click reaction in water. Nucleic Acids Res 35(21):e139. doi:10.1093/nar/gkm935
25. Brown SD, Graham D (2010) Conjugation of an oligonucleotide to Tat, a cell-penetrating peptide, via click chemistry. Tetrahedron Lett 51(38):5032–5034. doi:10.1016/j.tetlet.2010.07.101
26. El-Sagheer AH, Brown T (2010) New strategy for the synthesis of chemically modified RNA constructs exemplified by hairpin and hammerhead ribozymes. Proc Natl Acad Sci USA 107(35):15329–15334. doi:10.1073/pnas.1006447107
27. Frolow O, Endeward B, Schiemann O, Prisner TF, Engels JW (2008) Nitroxide spin labeled RNA for long range distance measurements by EPR–PELDOR. Nucleic Acids Symp Ser 52:153–154. doi:10.1093/nass/nrn078
28. Piton N, Mu Y, Stock G, Prisner TF, Schiemann O, Engels JW (2007) Base-specific spin-labeling of RNA for structure determination. Nucleic Acids Res 35(9):3128–3143. doi:10.1093/nar/gkm169
29. Piton N, Schiemann O, Mu Y, Stock G, Prisner T, Engels JW (2005) Synthesis of spin-labeled RNAs for long range distance measurements by peldor. Nucleosides Nucleotides Nucleic Acids 24(5–7):771–775
30. Schiemann O, Weber A, Edwards TE, Prisner TF, Sigurdsson ST (2003) Nanometer distance measurements on RNA using PELDOR. J Am Chem Soc 125(12):3434–3435. doi:10.1021/ja0274610
31. Ding P, Wunnicke D, Steinhoff HJ, Seela F (2010) Site-directed spin-labeling of DNA by the azide–alkyne 'click' reaction: nanometer distance measurements on 7-deaza-2′-deoxyadenosine and 2′-deoxyuridine nitroxide conjugates spatially separated or linked to a 'dA–dT' base pair. Chemistry 16(48):14385–14396. doi:10.1002/chem.201001572
32. Jakobsen U, Shelke SA, Vogel S, Sigurdsson ST (2010) Site-directed spin-labeling of nucleic acids by click chemistry: detection of abasic sites in duplex DNA by EPR spectroscopy. J Am Chem Soc 132(30):10424–10428. doi:10.1021/ja102797k
33. Wada T, Mochizuki A, Higashiya S, Tsuruoka H, S-i Kawahara, Ishikawa M, Sekine M (2001) Synthesis and properties of 2-azidodeoxyadenosine and its incorporation into oligodeoxynucleotides. Tetrahedron Lett 42(52):9215–9219. doi:10.1016/S0040-4039(01)02028-7
34. Pourceau G, Meyer A, Vasseur JJ, Morvan F (2009) Azide solid support for 3′-conjugation of oligonucleotides and their circularization by click chemistry. J Organ Chem 74(17):6837–6842. doi:10.1021/jo9014563
35. Neef AB, Luedtke NW (2014) An azide-modified nucleoside for metabolic labeling of DNA. Chembiochem 15(6):789–793. doi:10.1002/cbic.201400037
36. Kiviniemi A, Virta P, Lonnberg H (2008) Utilization of intrachain 4′-*C*-azidomethylthymidine for preparation of oligodeoxyribonucleotide conjugates by click chemistry in solution and on a solid support. Bioconjug Chem 19(8):1726–1734. doi:10.1021/bc800221p
37. Aigner M, Hartl M, Fauster K, Steger J, Bister K, Micura R (2011) Chemical synthesis of site-specifically 2′-azido-modified RNA and potential applications for bioconjugation and RNA interference. Chembiochem 12(1):47–51. doi:10.1002/cbic.201000646
38. Fauster K, Hartl M, Santner T, Aigner M, Kreutz C, Bister K, Ennifar E, Micura R (2012) 2′-Azido RNA, a versatile tool for chemical biology: synthesis, X-ray structure, siRNA applications, click labeling. ACS Chem Biol 7(3):581–589. doi:10.1021/cb200510k

39. Santner T, Hartl M, Bister K, Micura R (2014) Efficient access to 3′-terminal azide-modified RNA for inverse click-labeling patterns. Bioconjug Chem 25(1):188–195. doi:10.1021/bc400513z
40. Saxon E, Bertozzi CR (2000) Cell surface engineering by a modified Staudinger reaction. Science 287(5460):2007–2010
41. Wang CC, Seo TS, Li Z, Ruparel H, Ju J (2003) Site-specific fluorescent labeling of DNA using Staudinger ligation. Bioconjug Chem 14(3):697–701. doi:10.1021/bc0256392
42. Weisbrod SH, Baccaro A, Marx A (2008) DNA conjugation by Staudinger ligation. Nucleic Acids Symp Ser 52:383–384. doi:10.1093/nass/nrn195
43. Weisbrod SH, Baccaro A, Marx A (2011) Site-specific DNA labeling by Staudinger ligation. Methods Mol Biol 751:195–207. doi:10.1007/978-1-61779-151-2_12
44. Seela F, Pujari SS (2010) Azide–alkyne “click” conjugation of 8-aza-7-deazaadenine-DNA: synthesis, duplex stability, and fluorogenic dye labeling. Bioconjug Chem 21(9):1629–1641. doi:10.1021/bc100090y
45. Soriano Del Amo D, Wang W, Jiang H, Besanceney C, Yan AC, Levy M, Liu Y, Marlow FL, Wu P (2010) Biocompatible copper(I) catalysts for in vivo imaging of glycans. J Am Chem Soc 132(47):16893–16899. doi:10.1021/ja106553e
46. Kennedy DC, McKay CS, Legault MC, Danielson DC, Blake JA, Pegoraro AF, Stolow A, Mester Z, Pezacki JP (2011) Cellular consequences of copper complexes used to catalyze bioorthogonal click reactions. J Am Chem Soc 133(44):17993–18001. doi:10.1021/ja2083027
47. Eltepu L, Jayaraman M, Rajeev KG, Manoharan M (2013) An immobilized and reusable Cu(I) catalyst for metal ion-free conjugation of ligands to fully deprotected oligonucleotides through click reaction. Chem Commun 49(2):184–186. doi:10.1039/c2cc36811k
48. Jewett JC, Sletten EM, Bertozzi CR (2010) Rapid Cu-free click chemistry with readily synthesized biarylazacyclooctynones. J Am Chem Soc 132(11):3688–3690. doi:10.1021/ja100014q
49. Chang PV, Prescher JA, Sletten EM, Baskin JM, Miller IA, Agard NJ, Lo A, Bertozzi CR (2010) Copper-free click chemistry in living animals. Proc Natl Acad Sci USA 107(5):1821–1826. doi:10.1073/pnas.0911116107
50. van Delft P, Meeuwenoord NJ, Hoogendoorn S, Dinkelaar J, Overkleeft HS, van der Marel GA, Filippov DV (2010) Synthesis of oligoribonucleic acid conjugates using a cyclooctyne phosphoramidite. Org Lett 12(23):5486–5489. doi:10.1021/ol102357u
51. Singh I, Freeman C, Madder A, Vyle JS, Heaney F (2012) Fast RNA conjugations on solid phase by strain-promoted cycloadditions. Org Biomol Chem 10(33):6633–6639. doi:10.1039/c2ob25628b
52. Jayaprakash KN, Peng CG, Butler D, Varghese JP, Maier MA, Rajeev KG, Manoharan M (2010) Non-nucleoside building blocks for copper-assisted and copper-free click chemistry for the efficient synthesis of RNA conjugates. Org Lett 12(23):5410–5413. doi:10.1021/ol102205j
53. Shelbourne M, Chen X, Brown T, El-Sagheer AH (2011) Fast copper-free click DNA ligation by the ring-strain promoted alkyne–azide cycloaddition reaction. Chem Commun 47(22):6257–6259. doi:10.1039/c1cc10743g
54. Shelbourne M, Brown T Jr, El-Sagheer AH, Brown T (2012) Fast and efficient DNA crosslinking and multiple orthogonal labelling by copper-free click chemistry. Chem Commun 48(91):11184–11186. doi:10.1039/c2cc35084j
55. Marks IS, Kang JS, Jones BT, Landmark KJ, Cleland AJ, Taton TA (2011) Strain-promoted “click” chemistry for terminal labeling of DNA. Bioconjug Chem 22(7):1259–1263. doi:10.1021/bc1003668
56. Jawalekar AM, Malik S, Verkade JM, Gibson B, Barta NS, Hodges JC, Rowan A, van Delft FL (2013) Oligonucleotide tagging for copper-free click conjugation. Molecules 18(7):7346–7363. doi:10.3390/molecules18077346
57. Heuer-Jungemann A, Kirkwood R, El-Sagheer AH, Brown T, Kanaras AG (2013) Copper-free click chemistry as an emerging tool for the programmed ligation of DNA-functionalised gold nanoparticles. Nanoscale 5(16):7209–7212. doi:10.1039/c3nr02362a
58. Debets MF, van Berkel SS, Dommerholt J, Dirks AT, Rutjes FP, van Delft FL (2011) Bioconjugation with strained alkenes and alkynes. Acc Chem Res 44(9):805–815. doi:10.1021/ar200059z
59. Stubinitzky C, Cserep GB, Batzner E, Kele P, Wagenknecht HA (2014) 2′-Deoxyuridine conjugated with a reactive monobenzocyclooctyne as a DNA building block for copper-free click-type postsynthetic modification of DNA. Chem Commun 50(76):11218–11221. doi:10.1039/c4cc02855d
60. Gutsmiedl K, Wirges CT, Ehmke V, Carell T (2009) Copper-free “click” modification of DNA via nitrile oxide-norbornene 1,3-dipolar cycloaddition. Org Lett 11(11):2405–2408. doi:10.1021/ol9005322

61. Song W, Wang Y, Qu J, Lin Q (2008) Selective functionalization of a genetically encoded alkene-containing protein via "photoclick chemistry" in bacterial cells. J Am Chem Soc 130(30):9654–9655. doi:10.1021/ja803598e
62. Arndt S, Wagenknecht HA (2014) "Photoclick" postsynthetic modification of DNA. Angew Chem 53(52):14580–14582. doi:10.1002/anie.201407874
63. Sletten EM, Bertozzi CR (2009) Bioorthogonal chemistry: fishing for selectivity in a sea of functionality. Angew Chem 48(38):6974–6998. doi:10.1002/anie.200900942
64. Schoch J, Staudt M, Samanta A, Wiessler M, Jäschke A (2012) Site-specific one-pot dual labeling of DNA by orthogonal cycloaddition chemistry. Bioconjug Chem 23(7):1382–1386. doi:10.1021/bc300181n
65. Cole CM, Yang J, Šečkutė J, Devaraj NK (2013) Fluorescent live-cell imaging of metabolically incorporated unnatural cyclopropene–mannosamine derivatives. Chembiochem 14(2):205–208. doi:10.1002/cbic.201200719
66. Devaraj NK, Weissleder R, Hilderbrand SA (2008) Tetrazine-based cycloadditions: application to pretargeted live cell imaging. Bioconjug Chem 19(12):2297–2299. doi:10.1021/bc8004446
67. Devaraj NK, Upadhyay R, Haun JB, Hilderbrand SA, Weissleder R (2009) Fast and sensitive pretargeted labeling of cancer cells through a tetrazine/*trans*-cyclooctene cycloaddition. Angew Chem 48(38):7013–7016. doi:10.1002/anie.200903233
68. Devaraj NK, Weissleder R (2011) Biomedical applications of tetrazine cycloadditions. Acc Chem Res 44(9):816–827. doi:10.1021/ar200037t
69. Lang K, Davis L, Torres-Kolbus J, Chou C, Deiters A, Chin JW (2012) Genetically encoded norbornene directs site-specific cellular protein labelling via a rapid bioorthogonal reaction. Nat Chem 4(4):298–304. doi:10.1038/nchem.1250
70. Kaya E, Vrabel M, Deiml C, Prill S, Fluxa VS, Carell T (2012) A genetically encoded norbornene amino acid for the mild and selective modification of proteins in a copper-free click reaction. Angew Chem Int Ed Engl 51(18):4466–4469. doi:10.1002/anie.201109252
71. Liu DS, Tangpeerachaikul A, Selvaraj R, Taylor MT, Fox JM, Ting AY (2012) Diels–Alder cycloaddition for fluorophore targeting to specific proteins inside living cells. J Am Chem Soc 134(2):792–795. doi:10.1021/ja209325n
72. Blackman ML, Royzen M, Fox JM (2008) Tetrazine ligation: fast bioconjugation based on inverse-electron-demand Diels–Alder reactivity. J Am Chem Soc 130(41):13518–13519. doi:10.1021/ja8053805
73. Devaraj NK, Hilderbrand S, Upadhyay R, Mazitschek R, Weissleder R (2010) Bioorthogonal turn-on probes for imaging small molecules inside living cells. Angew Chem 49(16):2869–2872. doi:10.1002/anie.200906120
74. Carlson JC, Meimetis LG, Hilderbrand SA, Weissleder R (2013) BODIPY-tetrazine derivatives as superbright bioorthogonal turn-on probes. Angew Chem Int Ed Engl 52(27):6917–6920. doi:10.1002/anie.201301100
75. Karver MR, Weissleder R, Hilderbrand SA (2011) Synthesis and evaluation of a series of 1,2,4,5-tetrazines for bioorthogonal conjugation. Bioconjug Chem 22(11):2263–2270. doi:10.1021/bc200295y
76. Schoch J, Wiessler M, Jäschke A (2010) Post-synthetic modification of DNA by inverse-electron-demand Diels–Alder reaction. J Am Chem Soc 132(26):8846–8847. doi:10.1021/ja102871p
77. Šečkutė J, Yang J, Devaraj NK (2013) Rapid oligonucleotide-templated fluorogenic tetrazine ligations. Nucleic Acids Res 41(15):e148. doi:10.1093/nar/gkt540
78. Wang K, Wang D, Ji K, Chen W, Zheng Y, Dai C, Wang B (2014) Post-synthesis DNA modifications using a *trans*-cyclooctene click handle. Org Biomol Chem 13:909–915. doi:10.1039/C4OB02031F
79. Schoch J, Ameta S, Jäschke A (2011) Inverse electron-demand Diels–Alder reactions for the selective and efficient labeling of RNA. Chem Commun 47(46):12536–12537. doi:10.1039/c1cc15476a
80. Pyka AM, Domnick C, Braun F, Kath-Schorr S (2014) Diels–Alder cycloadditions on synthetic RNA in mammalian cells. Bioconjug Chem 25(8):1438–1443. doi:10.1021/bc500302y
81. Ameta S, Becker J, Jäschke A (2014) RNA-peptide conjugate synthesis by inverse-electron demand Diels–Alder reaction. Org Biomol Chem 12(26):4701–4707. doi:10.1039/c4ob00076e
82. Asare-Okai PN, Agustin E, Fabris D, Royzen M (2014) Site-specific fluorescence labelling of RNA using bio-orthogonal reaction of *trans*-cyclooctene and tetrazine. Chem Commun 50(58):7844–7847. doi:10.1039/c4cc02435d

83. Vranken C, Deen J, Dirix L, Stakenborg T, Dehaen W, Leen V, Hofkens J, Neely RK (2014) Super-resolution optical DNA Mapping via DNA methyltransferase-directed click chemistry. Nucleic Acids Res 42(7):e50. doi:10.1093/nar/gkt1406
84. Motorin Y, Burhenne J, Teimer R, Koynov K, Willnow S, Weinhold E, Helm M (2011) Expanding the chemical scope of RNA: methyltransferases to site-specific alkynylation of RNA for click labeling. Nucleic Acids Res 39(5):1943–1952. doi:10.1093/nar/gkq825
85. Tomkuviene M, Clouet-d'Orval B, Cerniauskas I, Weinhold E, Klimasauskas S (2012) Programmable sequence-specific click-labeling of RNA using archaeal box C/D RNP methyltransferases. Nucleic Acids Res 40(14):6765–6773. doi:10.1093/nar/gks381
86. Holstein JM, Stummer D, Rentmeister A (2015) Enzymatic modification of 5′-capped RNA with a 4-vinylbenzyl group provides a platform for photoclick and inverse electron-demand Diels–Alder reaction. Chem Sci 6(2):1362–1369. doi:10.1039/C4SC03182B
87. Silverman SK, Baum DA (2009) Use of deoxyribozymes in RNA research. Methods Enzymol 469:95–117. doi:10.1016/S0076-6879(09)69005-4
88. Büttner L, Javadi-Zarnaghi F, Höbartner C (2014) Site-specific labeling of RNA at internal ribose hydroxyl groups: terbium-assisted deoxyribozymes at work. J Am Chem Soc 136(22):8131–8137. doi:10.1021/ja503864v
89. El-Sagheer AH, Brown T (2012) Click nucleic acid ligation: applications in biology and nanotechnology. Acc Chem Res 45(8):1258–1267. doi:10.1021/ar200321n
90. Isobe H, Fujino T, Yamazaki N, Guillot-Nieckowski M, Nakamura E (2008) Triazole-linked analogue of deoxyribonucleic acid ((TL)DNA): design, synthesis, and double-strand formation with natural DNA. Org Lett 10(17):3729–3732. doi:10.1021/ol801230k
91. El-Sagheer AH, Brown T (2009) Synthesis and polymerase chain reaction amplification of DNA strands containing an unnatural triazole linkage. J Am Chem Soc 131(11):3958–3964. doi:10.1021/ja8065896
92. Qiu J, El-Sagheer AH, Brown T (2013) Solid phase click ligation for the synthesis of very long oligonucleotides. Chem Commun 49(62):6959–6961. doi:10.1039/c3cc42451k
93. El-Sagheer AH, Sanzone AP, Gao R, Tavassoli A, Brown T (2011) Biocompatible artificial DNA linker that is read through by DNA polymerases and is functional in *Escherichia coli*. Proc Natl Acad Sci USA 108(28):11338–11343. doi:10.1073/pnas.1101519108
94. Birts CN, Sanzone AP, El-Sagheer AH, Blaydes JP, Brown T, Tavassoli A (2014) Transcription of click-linked DNA in human cells. Angew Chem 53(9):2362–2365. doi:10.1002/anie.201308691
95. Chen X, El-Sagheer AH, Brown T (2014) Reverse transcription through a bulky triazole linkage in RNA: implications for RNA sequencing. Chem Commun 50(57):7597–7600. doi:10.1039/c4cc03027c
96. Stark MR, Pleiss JA, Deras M, Scaringe SA, Rader SD (2006) An RNA ligase-mediated method for the efficient creation of large, synthetic RNAs. RNA 12(11):2014–2019. doi:10.1261/rna.93506
97. Mattick JS, Makunin IV (2006) Non-coding RNA. Hum Mol Genet 15(Suppl 1):R17–R29. doi:10.1093/hmg/ddl046
98. Morris KV, Mattick JS (2014) The rise of regulatory RNA. Nat Rev Genet 15(6):423–437. doi:10.1038/nrg3722
99. Switzer C, Moroney SE, Benner SA (1989) Enzymatic incorporation of a new base pair into DNA and RNA. J Am Chem Soc 111(21):8322–8323. doi:10.1021/ja00203a067
100. Switzer CY, Moroney SE, Benner SA (1993) Enzymic recognition of the base pair between isocytidine and isoguanosine. Biochemistry 32(39):10489–10496. doi:10.1021/bi00090a027
101. Geyer CR, Battersby TR, Benner SA (2003) Nucleobase pairing in expanded Watson–Crick-like genetic information systems. Structure 11(12):1485–1498
102. Moser MJ, Marshall DJ, Grenier JK, Kieffer CD, Killeen AA, Ptacin JL, Richmond CS, Roesch EB, Scherrer CW, Sherrill CB, Van Hout CV, Zanton SJ, Prudent JR (2003) Exploiting the enzymatic recognition of an unnatural base pair to develop a universal genetic analysis system. Clin Chem 49(3):407–414
103. Johnson SC, Sherrill CB, Marshall DJ, Moser MJ, Prudent JR (2004) A third base pair for the polymerase chain reaction: inserting isoC and isoG. Nucleic Acids Res 32(6):1937–1941. doi:10.1093/nar/gkh522
104. Yang Z, Sismour AM, Sheng P, Puskar NL, Benner SA (2007) Enzymatic incorporation of a third nucleobase pair. Nucleic Acids Res 35(13):4238–4249. doi:10.1093/nar/gkm395
105. Hirao I, Mitsui T, Kimoto M, Yokoyama S (2007) Development of an unnatural base pair for efficient PCR amplification. Nucleic Acids Symp Ser 51:9–10. doi:10.1093/nass/nrm005

106. Kimoto M, Kawai R, Mitsui T, Yokoyama S, Hirao I (2009) An unnatural base pair system for efficient PCR amplification and functionalization of DNA molecules. Nucleic Acids Res 37(2):e14. doi:10.1093/nar/gkn956
107. Seo YJ, Hwang GT, Ordoukhanian P, Romesberg FE (2009) Optimization of an unnatural base pair toward natural-like replication. J Am Chem Soc 131(9):3246–3252. doi:10.1021/ja807853m
108. Yang Z, Chen F, Alvarado JB, Benner SA (2011) Amplification, mutation, and sequencing of a six-letter synthetic genetic system. J Am Chem Soc 133(38):15105–15112. doi:10.1021/ja204910n
109. Hirao I, Kimoto M (2012) Unnatural base pair systems toward the expansion of the genetic alphabet in the central dogma. Proc Jpn Acad Ser B Phys Biol Sci 88(7):345–367
110. Yamashige R, Kimoto M, Takezawa Y, Sato A, Mitsui T, Yokoyama S, Hirao I (2012) Highly specific unnatural base pair systems as a third base pair for PCR amplification. Nucleic Acids Res 40(6):2793–2806. doi:10.1093/nar/gkr1068
111. Li L, Degardin M, Lavergne T, Malyshev DA, Dhami K, Ordoukhanian P, Romesberg FE (2014) Natural-like replication of an unnatural base pair for the expansion of the genetic alphabet and biotechnology applications. J Am Chem Soc 136(3):826–829. doi:10.1021/ja408814g
112. Malyshev DA, Dhami K, Lavergne T, Chen T, Dai N, Foster JM, Correa IR Jr, Romesberg FE (2014) A semi-synthetic organism with an expanded genetic alphabet. Nature 509(7500):385–388. doi:10.1038/nature13314
113. Matray TJ, Kool ET (1998) Selective and stable DNA base pairing without hydrogen bonds. J Am Chem Soc 120(24):6191–6192. doi:10.1021/ja9803310
114. Matray TJ, Kool ET (1999) A specific partner for abasic damage in DNA. Nature 399(6737):704–708. doi:10.1038/21453
115. Morales JC, Kool ET (2000) Importance of terminal base pair hydrogen-bonding in 3′-end proofreading by the Klenow fragment of DNA polymerase I. Biochemistry 39(10):2626–2632
116. Betz K, Malyshev DA, Lavergne T, Welte W, Diederichs K, Dwyer TJ, Ordoukhanian P, Romesberg FE, Marx A (2012) KlenTaq polymerase replicates unnatural base pairs by inducing a Watson–Crick geometry. Nat Chem Biol 8(7):612–614. doi:10.1038/nchembio.966
117. Betz K, Malyshev DA, Lavergne T, Welte W, Diederichs K, Romesberg FE, Marx A (2013) Structural insights into DNA replication without hydrogen bonds. J Am Chem Soc 135(49):18637–18643. doi:10.1021/ja409609j
118. Dhami K, Malyshev DA, Ordoukhanian P, Kubelka T, Hocek M, Romesberg FE (2014) Systematic exploration of a class of hydrophobic unnatural base pairs yields multiple new candidates for the expansion of the genetic alphabet. Nucleic Acids Res 42(16):10235–10244. doi:10.1093/nar/gku715
119. Lavergne T, Degardin M, Malyshev DA, Quach HT, Dhami K, Ordoukhanian P, Romesberg FE (2013) Expanding the scope of replicable unnatural DNA: stepwise optimization of a predominantly hydrophobic base pair. J Am Chem Soc 135(14):5408–5419. doi:10.1021/ja312148q
120. Li Z, Lavergne T, Malyshev DA, Zimmermann J, Adhikary R, Dhami K, Ordoukhanian P, Sun Z, Xiang J, Romesberg FE (2013) Site-specifically arraying small molecules or proteins on DNA using an expanded genetic alphabet. Chemistry 19(42):14205–14209. doi:10.1002/chem.201302496
121. Malyshev DA, Dhami K, Quach HT, Lavergne T, Ordoukhanian P, Torkamani A, Romesberg FE (2012) Efficient and sequence-independent replication of DNA containing a third base pair establishes a functional six-letter genetic alphabet. Proc Natl Acad Sci USA 109(30):12005–12010. doi:10.1073/pnas.1205176109
122. Malyshev DA, Pfaff DA, Ippoliti SI, Hwang GT, Dwyer TJ, Romesberg FE (2010) Solution structure, mechanism of replication, and optimization of an unnatural base pair. Chemistry 16(42):12650–12659. doi:10.1002/chem.201000959
123. Malyshev DA, Seo YJ, Ordoukhanian P, Romesberg FE (2009) PCR with an expanded genetic alphabet. J Am Chem Soc 131(41):14620–14621. doi:10.1021/ja906186f
124. Endo M, Mitsui T, Okuni T, Kimoto M, Hirao I, Yokoyama S (2004) Unnatural base pairs mediate the site-specific incorporation of an unnatural hydrophobic component into RNA transcripts. Bioorg Med Chem Lett 14(10):2593–2596. doi:10.1016/j.bmcl.2004.02.072
125. Hikida Y, Kimoto M, Yokoyama S, Hirao I (2010) Site-specific fluorescent probing of RNA molecules by unnatural base-pair transcription for local structural conformation analysis. Nat Protoc 5(7):1312–1323. doi:10.1038/nprot.2010.77
126. Hirao I, Kimoto M, Yamashige R (2012) Natural versus artificial creation of base pairs in DNA: origin of nucleobases from the perspectives of unnatural base pair studies. Acc Chem Res 45(12):2055–2065. doi:10.1021/ar200257x

127. Hirao I, Mitsui T, Kimoto M, Kawai R, Sato A, Yokoyama S (2005) Non-hydrogen-bonded base pairs for specific transcription. Nucleic Acids Symp Ser 49:33–34. doi:10.1093/nass/49.1.33
128. Hirao I, Mitsui T, Kimoto M, Yokoyama S (2007) An efficient unnatural base pair for PCR amplification. J Am Chem Soc 129(50):15549–15555. doi:10.1021/ja073830m
129. Kimoto M, Kawai R, Mitsui T, Yokoyama S, Hirao I (2008) Efficient PCR amplification by an unnatural base pair system. Nucleic Acids Symp Ser 52:469–470. doi:10.1093/nass/nrn238
130. Kimoto M, Kawai R, Mitsui T, Yokoyama S, Hirao I (2008) Sequences around the unnatural base pair in DNA templates for efficient replication. Nucleic Acids Symp Ser 52:457–458. doi:10.1093/nass/nrn232
131. Kimoto M, Mitsui T, Yokoyama S, Hirao I (2010) A unique fluorescent base analogue for the expansion of the genetic alphabet. J Am Chem Soc 132(14):4988–4989. doi:10.1021/ja100806c
132. Mitsui T, Kimoto M, Harada Y, Sato A, Kitamura A, To T, Hirao I, Yokoyama S (2002) Enzymatic incorporation of an unnatural base pair between 4-propynyl-pyrrole-2-carbaldehyde and 9-methyl-imidazo [(4,5)-b]pyridine into nucleic acids. Nucleic Acids Res Suppl 2:219–220
133. Mitsui T, Kimoto M, Harada Y, Yokoyama S, Hirao I (2005) An efficient unnatural base pair for a base-pair-expanded transcription system. J Am Chem Soc 127(24):8652–8658. doi:10.1021/ja0425280
134. Ohtsuki T, Kimoto M, Ishikawa M, Mitsui T, Hirao I, Yokoyama S (2001) Unnatural base pairs for specific transcription. Proc Natl Acad Sci USA 98(9):4922–4925. doi:10.1073/pnas.091532698
135. Ishizuka T, Kimoto M, Sato A, Hirao I (2012) Site-specific functionalization of RNA molecules by an unnatural base pair transcription system via click chemistry. Chem Commun 48(88):10835–10837. doi:10.1039/c2cc36293g
136. Kawai R, Kimoto M, Ikeda S, Mitsui T, Endo M, Yokoyama S, Hirao I (2005) Site-specific fluorescent labeling of RNA molecules by specific transcription using unnatural base pairs. J Am Chem Soc 127(49):17286–17295. doi:10.1021/ja0542946
137. Kawai R, Kimoto M, Mitsui T, Yokoyama S, Hirao I (2004) Site-specific fluorescent labeling of RNA by a base-pair expanded transcription system. Nucleic Acids Symp Ser 48:35–36. doi:10.1093/nass/48.1.35
138. Kimoto M, Hirao I (2010) Site-specific incorporation of extra components into RNA by transcription using unnatural base pair systems. Methods Mol Biol 634:355–369. doi:10.1007/978-1-60761-652-8_25
139. Kimoto M, Kawai R, Mitsui T, Harada Y, Sato A, Yokoyama S, Hirao I (2005) Site-specific incorporation of fluorescent probes into RNA by specific transcription using unnatural base pairs. Nucleic Acids Symp Ser 49:287–288. doi:10.1093/nass/49.1.287
140. Kimoto M, Sato A, Kawai R, Yokoyama S, Hirao I (2009) Site-specific incorporation of functional components into RNA by transcription using unnatural base pair systems. Nucleic Acids Symp Ser 53:73–74. doi:10.1093/nass/nrp037
141. Kimoto M, Yamashige R, Yokoyama S, Hirao I (2012) PCR amplification and transcription for site-specific labeling of large RNA molecules by a two-unnatural-base-pair system. J Nucleic Acids 2012:230943. doi:10.1155/2012/230943
142. Seo YJ, Matsuda S, Romesberg FE (2009) Transcription of an expanded genetic alphabet. J Am Chem Soc 131(14):5046–5047. doi:10.1021/ja9006996
143. Seo YJ, Malyshev DA, Lavergne T, Ordoukhanian P, Romesberg FE (2011) Site-specific labeling of DNA and RNA using an efficiently replicated and transcribed class of unnatural base pairs. J Am Chem Soc 133(49):19878–19888. doi:10.1021/ja207907d
144. Domnick C, Eggert F, Kath-Schorr S (2015) Site-specific enzymatic introduction of a norbornene modified unnatural base into RNA and application in post-transcriptional labeling. Chem Commun 51(39):8253–8256. doi:10.1039/c5cc01765c
145. Moriyama K, Kimoto M, Mitsui T, Yokoyama S, Hirao I (2005) Site-specific biotinylation of RNA molecules by transcription using unnatural base pairs. Nucleic Acids Res 33(15):e129. doi:10.1093/nar/gni128
146. Morohashi N, Kimoto M, Sato A, Kawai R, Hirao I (2012) Site-specific incorporation of functional components into RNA by an unnatural base pair transcription system. Molecules 17(3):2855–2876. doi:10.3390/molecules17032855
147. Li Z, Cai H, Hassink M, Blackman ML, Brown RC, Conti PS, Fox JM (2010) Tetrazine-*trans*-cyclooctene ligation for the rapid construction of 18F labeled probes. Chem Commun 46(42):8043–8045. doi:10.1039/c0cc03078c

148. Seitchik JL, Peeler JC, Taylor MT, Blackman ML, Rhoads TW, Cooley RB, Refakis C, Fox JM, Mehl RA (2012) Genetically encoded tetrazine amino acid directs rapid site-specific in vivo bioorthogonal ligation with *trans*-cyclooctenes. J Am Chem Soc 134(6):2898–2901. doi:10.1021/ja2109745
149. Schneider S, Gattner MJ, Vrabel M, Flügel V, Lopez-Carrillo V, Prill S, Carell T (2013) Structural insights into incorporation of norbornene amino acids for click modification of proteins. Chembiochem. doi:10.1002/cbic.201300435
150. Schulz D, Rentmeister A (2014) Current approaches for RNA labeling in vitro and in cells based on click reactions. Chembiochem 15(16):2342–2347. doi:10.1002/cbic.201402240
151. Chadalavada DM, Gratton EA, Bevilacqua PC (2010) The human HDV-like CPEB3 ribozyme is intrinsically fast-reacting. Biochemistry 49(25):5321–5330. doi:10.1021/bi100434c
152. Someya T, Ando A, Kimoto M, Hirao I (2015) Site-specific labeling of RNA by combining genetic alphabet expansion transcription and copper-free click chemistry. Nucleic Acids Res. doi:10.1093/nar/gkv638
153. Salic A, Mitchison TJ (2008) A chemical method for fast and sensitive detection of DNA synthesis in vivo. Proc Natl Acad Sci USA 105(7):2415–2420. doi:10.1073/pnas.0712168105
154. Chehrehasa F, Meedeniya AC, Dwyer P, Abrahamsen G, Mackay-Sim A (2009) EdU, a new thymidine analogue for labelling proliferating cells in the nervous system. J Neurosci Methods 177(1):122–130. doi:10.1016/j.jneumeth.2008.10.006
155. Zeng C, Pan F, Jones LA, Lim MM, Griffin EA, Sheline YI, Mintun MA, Holtzman DM, Mach RH (2010) Evaluation of 5-ethynyl-2′-deoxyuridine staining as a sensitive and reliable method for studying cell proliferation in the adult nervous system. Brain Res 1319:21–32. doi:10.1016/j.brainres.2009.12.092
156. Jao CY, Salic A (2008) Exploring RNA transcription and turnover in vivo by using click chemistry. Proc Natl Acad Sci USA 105(41):15779–15784. doi:10.1073/pnas.0808480105
157. Diermeier-Daucher S, Clarke ST, Hill D, Vollmann-Zwerenz A, Bradford JA, Brockhoff G (2009) Cell type specific applicability of 5-ethynyl-2′-deoxyuridine (EdU) for dynamic proliferation assessment in flow cytometry. Cytometry Part A J Int Soc Anal Cytol 75(6):535–546. doi:10.1002/cyto.a.20712
158. Ross HH, Rahman M, Levkoff LH, Millette S, Martin-Carreras T, Dunbar EM, Reynolds BA, Laywell ED (2011) Ethynyldeoxyuridine (EdU) suppresses in vitro population expansion and in vivo tumor progression of human glioblastoma cells. J Neurooncol 105(3):485–498. doi:10.1007/s11060-011-0621-6
159. Zhao H, Halicka HD, Li J, Biela E, Berniak K, Dobrucki J, Darzynkiewicz Z (2013) DNA damage signaling, impairment of cell cycle progression, and apoptosis triggered by 5-ethynyl-2′-deoxyuridine incorporated into DNA. Cytometry Part A J Int Soc Anal Cytol 83(11):979–988. doi:10.1002/cyto.a.22396
160. Neef AB, Luedtke NW (2011) Dynamic metabolic labeling of DNA in vivo with arabinosyl nucleosides. Proc Natl Acad Sci USA 108(51):20404–20409. doi:10.1073/pnas.1101126108
161. Qu D, Wang G, Wang Z, Zhou L, Chi W, Cong S, Ren X, Liang P, Zhang B (2011) 5-Ethynyl-2′-deoxycytidine as a new agent for DNA labeling: detection of proliferating cells. Anal Biochem 417(1):112–121. doi:10.1016/j.ab.2011.05.037
162. Guan L, van der Heijden GW, Bortvin A, Greenberg MM (2011) Intracellular detection of cytosine incorporation in genomic DNA by using 5-ethynyl-2′-deoxycytidine. Chembiochem 12(14):2184–2190. doi:10.1002/cbic.201100353
163. Neef AB, Samain F, Luedtke NW (2012) Metabolic labeling of DNA by purine analogues in vivo. Chembiochem 13(12):1750–1753. doi:10.1002/cbic.201200253
164. Wang IH, Suomalainen M, Andriasyan V, Kilcher S, Mercer J, Neef A, Luedtke NW, Greber UF (2013) Tracking viral genomes in host cells at single-molecule resolution. Cell Host Microbe 14(4):468–480. doi:10.1016/j.chom.2013.09.004
165. Hagemeijer MC, Vonk AM, Monastyrska I, Rottier PJ, de Haan CA (2012) Visualizing coronavirus RNA synthesis in time by using click chemistry. J Virol 86(10):5808–5816. doi:10.1128/JVI.07207-11
166. Neef AB, Pernot L, Schreier VN, Scapozza L, Luedtke NW (2015) A bioorthogonal chemical reporter of viral infection. Angew Chem 54(27):7911–7914. doi:10.1002/anie.201500250
167. Rieder U, Luedtke NW (2014) Alkene–tetrazine ligation for imaging cellular DNA. Angew Chem 53(35):9168–9172. doi:10.1002/anie.201403580

168. Sawant AA, Tanpure AA, Mukherjee PP, Athavale S, Kelkar A, Galande S, Srivatsan SG (2015) A versatile toolbox for posttranscriptional chemical labeling and imaging of RNA. Nucleic Acids Res. doi:10.1093/nar/gkv903
169. Klug SJ, Famulok M (1994) All you wanted to know about SELEX. Mol Biol Rep 20(2):97–107
170. Famulok M, Mayer G (2014) Aptamers and SELEX in chemistry and biology. Chem Biol 21(9):1055–1058. doi:10.1016/j.chembiol.2014.08.003
171. Kimoto M, Yamashige R, Matsunaga K, Yokoyama S, Hirao I (2013) Generation of high-affinity DNA aptamers using an expanded genetic alphabet. Nat Biotechnol 31(5):453–457. doi:10.1038/nbt.2556
172. Pinheiro VB, Taylor AI, Cozens C, Abramov M, Renders M, Zhang S, Chaput JC, Wengel J, Peak-Chew SY, McLaughlin SH, Herdewijn P, Holliger P (2012) Synthetic genetic polymers capable of heredity and evolution. Science 336(6079):341–344. doi:10.1126/science.1217622
173. Vaught JD, Bock C, Carter J, Fitzwater T, Otis M, Schneider D, Rolando J, Waugh S, Wilcox SK, Eaton BE (2010) Expanding the chemistry of DNA for in vitro selection. J Am Chem Soc 132(12):4141–4151. doi:10.1021/ja908035g
174. Sefah K, Yang Z, Bradley KM, Hoshika S, Jimenez E, Zhang L, Zhu G, Shanker S, Yu F, Turek D, Tan W, Benner SA (2014) In vitro selection with artificial expanded genetic information systems. Proc Natl Acad Sci USA 111(4):1449–1454. doi:10.1073/pnas.1311778111
175. Tolle F, Brandle GM, Matzner D, Mayer G (2015) A Versatile Approach Towards Nucleobase-Modified Aptamers. Angew Chem. doi:10.1002/anie.201503652
176. Lundberg EP, El-Sagheer AH, Kocalka P, Wilhelmsson LM, Brown T, Norden B (2010) A new fixation strategy for addressable nano-network building blocks. Chem Commun 46(21):3714–3716. doi:10.1039/c001513j
177. Qing G, Xiong H, Seela F, Sun T (2010) Spatially controlled DNA nanopatterns by "click" chemistry using oligonucleotides with different anchoring sites. J Am Chem Soc 132(43):15228–15232. doi:10.1021/ja105246b
178. Xiong H, Leonard P, Seela F (2012) Construction and assembly of branched Y-shaped DNA: "click" chemistry performed on dendronized 8-aza-7-deazaguanine oligonucleotides. Bioconjug Chem 23(4):856–870. doi:10.1021/bc300013k
179. Gerrard SR, Hardiman C, Shelbourne M, Nandhakumar I, Norden B, Brown T (2012) A new modular approach to nanoassembly: stable and addressable DNA nanoconstructs via orthogonal click chemistries. ACS Nano 6(10):9221–9228. doi:10.1021/nn3035759
180. Cassinelli V, Oberleitner B, Sobotta J, Nickels P, Grossi G, Kempter S, Frischmuth T, Liedl T, Manetto A (2015) One-step formation of "chain-armor"-stabilized DNA nanostructures. Angew Chem 54(27):7795–7798. doi:10.1002/anie.201500561

The manufacturer's authorised representative in the EU is Springer Nature Customer Service Centre GmbH, Europaplatz 3, 69115 Heidelberg, Germany. If you have any concerns regarding our products, please contact ProductSafety@springernature.com

Printed and bound by CPI Group (UK) Ltd, Croydon, CR0 4YY
25/09/2026
02222850-0006